On Beyond Uranium

Science Spectra Book Series

Series Editor: Vivian Moses, King's College, London, UK

This book is part of a series. The publisher will accept continuation orders which may be cancelled at any time and which provide for automatic billing and shipping of each title in the series upon publication. Please write for details.

On Beyond Uranium

Journey to the end of the Periodic Table

Sigurd Hofmann

London and New York

First published 2002
by Taylor & Francis
11 New Fetter lane, London EC4P 4EE

Simultaneously published in the USA and Canada
by Taylow & Francis Inc,
29 West 35th Street, New York, NY 10001

Taylor & Francis is an imprint of the Taylor & Francis Group

© 2002 Taylor & Francis

Typeset in Optima by Wearset Ltd, Boldon, Tyne and Wear
Printed and bound in Malta by Gutenberg Press Ltd

Every effort has been made to ensure that the advice and information in this book is true and accurate at the time of going to press. However, neither the publisher nor the authors can accept any legal responsibility or liability for any errors or omissions that may be made. In the case of drug administation, any medical procedure or the use of technical equipment mentioned within this book, you are strongly advised to consult the manufacturer's guidelines.

Every attempt has been made to obtain permission to reproduce copyright material. If any proper acknowledgement has not been made, we would invite copyright holders to inform us of this oversight.

British Library Cataloguing in Publication Data
A catalogue record for this book is available from the British Library

Library of Congress Cataloging in Publication Data
A catalog record for this book has been requested

ISBN 0-415-28495-3 (hbk)
ISBN 0-415-28496-1 (pbk)

Contents

Preface

The nineteenth-century chemists did a marvellous job. From no more than a primitive understanding of chemistry and the nature of the elements in 1800, ten decades later chemical science knew of the probable existence of 92 elements, from hydrogen the lightest to uranium the heaviest (although a few were so rare that they had not yet actually been identified). For nigh on 40 years we thought we knew what the Earth and everything on it and in it was made of.

Then things started to change. Towards the end of the 1930s chemists and physicists began tentatively to discover elements beyond uranium, the 'transuranics'. After World War II, with its startling developments in atomic physics leading both to atomic bombs and nuclear power stations, the pace began to quicken. When I started to study physics at the University of Darmstadt in 1963, there were 103 elements known; now there are 116 and the number is still rising, albeit slowly. Looking for new elements is one of the strangest treasure hunts ever.

You don't find them in the rocks or in the sea or even in the stars. Most probably they really are not found in Nature at all except, perhaps, in a most transitory sense – here one moment, gone the next. In the sense that they do 'exist', they are to be found only in very carefully and precisely designed experiments in powerful nuclear tools of various sorts. Everything has to be set up all ready to detect them the instant they are formed because, for some of them, their survival time is to be measured in fractions of a second.

Looking for new superheavy elements is what I spend my time doing – and I cannot imagine a more exciting branch of science. It is every bit as enthralling as it once was to explore an unknown continent but without mosquitoes and hard camp beds. This book is about the search for elements never before known to humankind. The story takes a bit of telling; some of the concepts will be unfamiliar and the terminology new for you. But stick with it through the more difficult parts; it is worth staying the course until the end to find out why I remain so enthusiastic!

Acknowledgements

Extracts on pages 28, 31, 32, 33, 34, 39, 40, 41, 42 are taken from Seaborg, G.T. and Loveland, W.D. (1990) *The elements beyond uranium.* John Wiley and Sons, Inc., New York. This material is used by permission of John Wiley and Sons, Inc.

The landscape

Our world of elements is a landscape of minute particles, of protons and neutrons, of electrons and their counterpart, neutrinos. From that set of four sub-atomic particles a whole world is made, a world of planets and mountains, of rivers, lakes, shores and oceans, of flowers and trees and animals – and also of ourselves, we the thinking people.

How is this possible – four tiny pieces and such a variety of consequences? Because the particles carry three secrets: the forces acting between them (which determine how the particles attract and repel each other, how they combine, and how they separate), the radiation which can be emitted from them when they move (this is the way we get the light from our sun), and the huge numbers of bricks available and how tiny each one is.

Nowadays, many people have some understanding of chemistry and know about the existence of atoms and protons, and neutrons and electrons, although neutrinos remain something for specialists. Chemistry grew from *alchemy*; the alchemists, who, you will remember, were forever looking for the philosopher's stone which would turn base metals into gold, with which their rulers never seemed to become sated. One of those alchemists, Johann Friedrich Böttger of Dresden, found porcelain instead of gold and was kept a prisoner in his laboratory so as not to betray the secret. All that looking at natural substances and experimenting with the ways in which they interacted with one another did, in the course of time, throw up a vast amount of empirical information which eventually demanded explanation. So chemistry became a modern experimental science, unlike the attempt at magic which had been the basis of the alchemy which had gone before. At about the same time, in the seventeenth century, the quality of astronomy reached a high standard. Sir Isaac Newton identified the universal law of gravitational force. Light and gases were investigated and answers sought to such apparently simple questions as the nature of air.

The nineteenth century was the century of *electromagnetic force*, the force which acts between the charges of the fundamental particles. Michael Faraday in London coined the term *ion* for charged particles moving in an electrolyte (an electrically conducting solution) between the electrodes. James Clerk Maxwell in Cambridge developed his theory of electromagnetism in 1864 and, based on that theory, Heinrich Hertz in Karlsruhe was able in 1887 to generate and measure radio waves. His name is used today in the symbol *Hz*, indicating the number of oscillations (periodic vibrations) per second.

Empirical chemistry reached its brilliant climax when, in 1860, the Periodic Table of the elements was introduced, proposed independently by the Russian Dmitri Ivanovitch Mendeleev in St Petersburg and the German Julius Lothar Meyer in Eberswalde, north of Berlin, and later in Karlsruhe. The Periodic Table ranks the elements based on their chemical behaviour and one of its most interesting and productive aspects was the presence of gaps which suggested the existence of elements which had not yet been discovered. Because the gaps occurred within an ordered sequence, Meyer and Mendeleev realised that there should be missing elements with chemical characteristics similar to those of known elements, such as aluminium and silicon. Knowing roughly what to look for made identification possible and, as early as 1875, gallium was discovered (closely resembling aluminium, as predicted) with germanium (related to silicon) following in 1886.

The complete list of the elements known to date is given in Table 1.1. The arrangement of the elements according to their chemical relationships is shown in the Periodic Table, Figure 1.1.

By the end of the nineteenth century, the ground had been prepared for the atomic age and the discovery of the elementary particles. In 1886, Eugen Goldstein in Berlin discovered *channel rays*, (low-energy heavy-ion beams) in gas discharges, while *X-rays* were discovered in 1895 by Wilhelm Conrad Röntgen in Würzburg and rapidly found wide application in medicine and technology. Note that, in the context of particle beams, 'energy' is related to speed; for a particular type of particle, the higher its energy, the faster it is moving. Only light and X-rays, which are both similar in principle, always move with the constant *velocity of light*.

In 1895, Antoine Henri Becquerel in Paris discovered something new: that rays could be emitted directly from uranium, a metal known to exist in certain naturally-occurring ores. The atomic nucleus had announced its existence but it took another 15 years before it could be observed directly. Still another kind of radiation was found in 1899 by Ernest Rutherford in Montreal, one which penetrated more deeply into material; he named these β-*rays* (beta-rays) to distinguish them from the less penetrating variety which he termed α-*rays* (alpha-rays).

In 1908, Hans Geiger and Ernest Marsden employed the α-rays to irradiate thin sheets of gold foil. This was done in Sir Ernest Rutherford's laboratory, now in Manchester. Using scintillators and a microscope, they observed

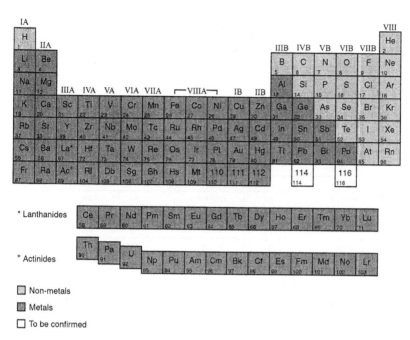

Figure 1.1 Periodic Table of the elements. The known transactinide elements 104 to 116 take the positions from below Hf in group IVA to below Po in group VIB. Element 108, hassium (Hs), the heaviest element chemically investigated, is placed in group VIIIA. The arrangement of the actinides reflects the fact that the first actinide elements still resemble, to a decreasing extent, the chemistry of the other groups: Th (group IVA below Hf), Pa (group VA below Ta) and U (group VIA below W).

unusual scattering angles of the α-particles. The explanation of these data given in 1911 by Rutherford was the real breakthrough into the atomic age: an atom consists of a tiny, heavy nucleus which is surrounded by electrons. The nucleus carries almost the whole mass of the atom, is positively charged and attracts the negatively charged electrons by electric forces.

By 1913, only two years later, Niels Hendrick David Bohr had, in Copenhagen, developed his atomic model which explained exactly the movement of the electrons around the nucleus in specific orbits characterised by certain energy and quantum numbers. Albert Einstein's famous 1905 law ($E = mc^2$)*, discovered when he was employed at the Bern patent

---

*Energy equals mass times the square of the velocity of light.

Table 1.1 The elements ordered by their atomic number

1	H	Hydrogen	41	Nb	Niobium	81	Tl	Thallium
2	He	Helium	42	Mo	Molybdenum	82	Pb	Lead
3	Li	Lithium	43	Tc	Technetium	83	Bi	Bismuth
4	Be	Beryllium	44	Ru	Ruthenium	84	Po	Polonium
5	B	Boron	45	Rh	Rhodium	85	At	Astatine
6	C	Carbon	46	Pd	Palladium	86	Rn	Radon
7	N	Nitrogen	47	Ag	Silver	87	Fr	Francium
8	O	Oxygen	48	Cd	Cadmium	88	Ra	Radium
9	F	Fluorine	49	In	Indium	89	Ac	Actinium
10	Ne	Neon	50	Sn	Tin	90	Th	Thorium
11	Na	Sodium	51	Sb	Antimony	91	Pa	Protactinium
12	Mg	Magnesium	52	Te	Tellurium	92	U	Uranium
13	Al	Aluminium	53	I	Iodine	93	Np	Neptunium
14	Si	Silicon	54	Xe	Xenon	94	Pu	Plutonium
15	P	Phosphorus	55	Cs	Caesium	95	Am	Americium
16	S	Sulphur	56	Ba	Barium	96	Cm	Curium
17	Cl	Chlorine	57	La	Lanthanum	97	Bk	Berkelium
18	Ar	Argon	58	Ce	Cerium	98	Cf	Californium
19	K	Potassium	59	Pr	Praseodymium	99	Es	Einsteinium
20	Ca	Calcium	60	Nd	Neodymium	100	Fm	Fermium
21	Sc	Scandium	61	Pm	Promethium	101	Md	Mendelevium
22	Ti	Titanium	62	Sm	Samarium	102	No	Nobelium
23	V	Vanadium	63	Eu	Europium	103	Lr	Lawrencium
24	Cr	Chromium	64	Gd	Gadolinium	104	Rf	Rutherfordium
25	Mn	Manganese	65	Tb	Terbium	105	Db	Dubnium
26	Fe	Iron	66	Dy	Dysprosium	106	Sg	Seaborgium
27	Co	Cobalt	67	Ho	Holmium	107	Bh	Bohrium
28	Ni	Nickel	68	Er	Erbium	108	Hs	Hassium
29	Cu	Copper	69	Tm	Thulium	109	Mt	Meitnerium
30	Zn	Zinc	70	Yb	Ytterbium	110		Name proposed
31	Ga	Gallium	71	Lu	Lutetium	111		Name proposed
32	Ge	Germanium	72	Hf	Hafnium	112		Name proposed
33	As	Arsenic	73	Ta	Tantalum	113		Unknown
34	Se	Selenium	74	W	Tungsten	114		To be confirmed
35	Br	Bromine	75	Re	Rhenium	115		Unknown
36	Kr	Krypton	76	Os	Osmium	116		To be confirmed
37	Rb	Rubidium	77	Ir	Iridium	117		Unknown
38	Sr	Strontium	78	Pt	Platinum	118		Unknown
39	Y	Yttrium	79	Au	Gold	119		Unknown
40	Zr	Zirconium	80	Hg	Mercury	120		Unknown

office, made possible the explanation of all of the discontinuous and continuous phenomena connected with emission and absorption of light and electrons. It was a phenomenally major advance.

High energy particles and atomic structure

The first experimental nuclear reaction, involving the collision between atoms and fast-moving highly energetic particles, took place in 1919 in Rutherford's laboratory. Although the scientists were using these experiments to explore the structure of atoms, it is easier to understand the significance of what they did if we take a brief look at what we now know.

Rutherford proposed that, like the solar system, atoms are mainly empty. At the centre of each one lies a small and very dense *nucleus* consisting of a number of positively charged *protons* (of a weight defined as 1 unit) together with a number of neutrons, also of weight 1 but uncharged. The weight of a nucleus equals the number of protons plus the number of neutrons and its net positive charge is the same as the number of protons. In orbits round the nucleus (just like planets round the sun) are electrons, each of which weighs almost nothing but carries a negative charge exactly cancelling the positive charge of a proton. The number of circulating electrons equals the net positive charge on the nucleus, so the atom as a whole is electrically neutral (see Box 1.1).

The *chemical* properties of an atom (i.e. what makes carbon different from nitrogen, or gold different from lead) depends on the number and configuration of the electrons, exactly which orbits they occupy and how easily they can enter and leave those orbits. However, an atom's physical properties (in particular how stable it is and whether it is likely to decay by the radioactive emission of a high-energy particle) is governed by the particular combination of protons and neutrons. Take the element carbon. Its nucleus contains six protons giving it a charge of +6; there are accordingly six electrons and that is what makes it behave chemically like carbon. Most carbon atoms in nature also have six neutrons; the resulting body weights 12 units and is perfectly stable. (It is actually the weight of this carbon atom divided by 12 which serves as our standard mass unit.) A second variant (*isotope* is the technical name) has seven neutrons and thus weights 13 units; it, too, is stable and, *because its nucleus still has a charge of +6*, it is chemically just as much carbon as the first. But a third variant has eight neutrons (its total weight is thus 14) and is unstable: it tends to decay by emitting a high-speed electron (β-ray) which means that one of its neutrons turns into a proton. So now there are seven protons and the nucleus has a charge of +7 and another electron will be picked up from the environment to balance the additional positive charge. But a nuclear charge of +7, balanced by seven electrons, is

Box 1.1 The Bohr atomic model

In the Bohr model and its later development, a small but dense nucleus lies at the centre of each atom; so dense is it that it accounts for more than 99.9% of the weight of an atom. It comprises one or more positively charged protons (each defined as having a unit weight = 1) plus a number of uncharged neutrons (each weighing only 1.4 per mill more than a proton), ranging from 0 in the very lightest atom to more than 150 in the heaviest.

Around the nucleus are arranged the negatively-charged electrons; the charge on an electron is exactly equal and opposite to that on a proton but its weight is nearly 2,000 times less. Since atoms are electrically neutral, the number of orbiting electrons exactly equals the number of protons in the nucleus. Thus, for hydrogen, the simplest of all atoms, one electron orbits around a nucleus of one proton.

The orbits expand in size the further they lie from the nucleus. The innermost can accommodate no more than two electrons, the next eight and so on up. The arrangement is most stable when an orbit (or 'electron shell') is full; atoms 'try' to fill or get rid of their outermost electron shells by sharing electrons with other atoms. Thus, when hydrogen shares its own electron with another atom, and in turn shares one *from* its partner, its own orbit fills up to the limit of 2. That sharing is a form of chemical bond between atoms. An example is the hydrogen molecule H_2. In some cases, as in our well known cooking salt sodium chloride (NaCl), the one atom (Na) gets more positively and the other (Cl) more negatively charged, because the only one outer electron of sodium 'likes' to move over to the partner chlorine and to fill up its outer electron shell, thus a net electrical attraction is obtained.

Of interest is a property of elementary particles, like the electrons, how they occupy a tiny volume, here the surroundings of two neighbouring atoms. At any time at each location within the volume there exists a certain probability which can be determined very accurately that one or two electrons should be there, but we never can say with certainty that the electron will be there or will be somewhere else. With certainty we can only say 'it is' somewhere within the volume. This is similar to the behaviour of railway trains in many countries. We never can say with certainty that there will be a train at a certain station at a certain time. Trains often behave like quantum-mechanical objects.

There is an atom which naturally has two electrons in its innermost shell, balancing a nucleus of two protons plus some neutrons: that element is helium. Because its electron shell is already full, it does not need to share and indeed forms chemical bonds only with great difficulty. Similar atoms (neon, argon, krypton, xenon and radon) exist with completed outer shells; they are collectively called *noble gases* because they are so reluctant to form compounds with the hoi polloi of the other elements! Element 118 is probably the next noble gas in this series (see Figure 1.1).

A few elements up from hydrogen is carbon with a nucleus containing six protons and (in various varieties or 'isotopes') 3, 4, 5, 6, 7, 8 up to 16 neutrons. A nucleus with six protons demands balancing with six electrons, two of them in the innermost shell (now full) and four in the next shell out which can hold a total of eight. Carbon tends to fill that outer shell by sharing four pairs of electrons, the four electrons of its own in that shell and four from partners. It can do so, for example, with four hydrogen atoms to make methane (CH_4), a very stable gas. Or it can spend its four electrons to two oxygen atoms (each of which has six electrons in its outer shell and so needs another two to fill the shell) to make carbon dioxide (CO_2), another stable compound. The number of sharings an atom can make determines much of its chemistry; there is an enormous number of such possible combinations.

Progressively as we go up the sequence with bigger and bigger (and heavier and heavier) atoms, the outer shells become larger. Then a new phenomenon appears: new outer shells are started without an inner one filling. It is only with some elements higher up the sequence that increasing nuclear charge starts being balanced by electrons *in that unfilled shell*, the outermost shell not changing for some 15 elements. Each one of those 15, therefore, has very similar *chemical* properties (determined by the number of electrons in their outermost shell) although they are getting heavier and heavier.

The element at which such a filling-up sequence begins is called *lanthanum* (with proton number $Z = 57$) and the series is therefore called the *lanthanides* (another name is 'rare earths' because they were first found in certain uncommon ores). As superheavy artificial atoms were made, one of the exciting facts which will emerge in our main story is that another such sequence begins at the element actinium ($Z = 89$) to give a series called the *actinides*.

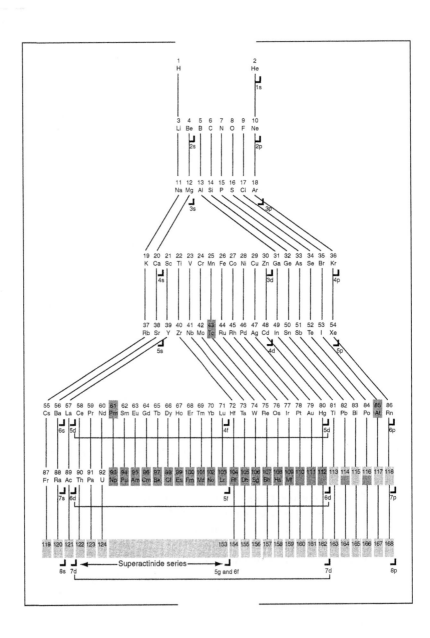

Exactly which orbits are filled and how the lanthanides and actinides fit into the Periodic Table are shown in Figure 1.2. There, the periodic arrangement of the elements is extended far beyond element 118 which is predicted to be a noble gas. The most striking feature of this arrangement is the addition of another family-like inner transition series starting at about atomic number 122 and extending through atomic number 153. The electron-shell structures predicted for the superheavy elements 114 and 126 are given schematically in Figure 1.3.

Figure 1.2 Structure and extension of the Periodic Table. The figure shows the natural and synthetic elements presently known up to element 116 together with hypothetical positions for additional elements through to 168. The 23 definitely known synthetic elements are indicated by rectangles in dark grey. The two rectangles at 114 and 116 mark elements which were reported and are now awaiting confirmation. Light grey rectangles indicate the synthetic elements not yet discovered. Most elements in each horizontal row differ from one another in chemical properties. The lines running from top to bottom connect elements of similar chemical properties. Above the symbol for each element is its atomic number: the number of positive charges in its nucleus or the number of electrons bound by them. In each horizontal row the bold brackets designated *1s, 2s, 2p* and so on denote the filling of subshells of electrons and it is largely the number of electrons in the outer shell which determines the chemical properties. In most elements, all the inner subshells are filled and electrons add to the outer shell with increasing atomic number. In the 'lanthanide' rare-earth elements (numbers 58 through to 71) the number of *5d* and *6s* electrons remains approximately the same and electrons in successive elements are added to the inner *4f* subshell. The 'actinide' elements (numbers 90 to 103) are part of another group of rare earths, in which the inner *5f* subshell fills up. A similar third series, the 'superactinides' was predicted by Glenn T. Seaborg in 1969 to run from about element 122 through to element 153. This series is formed by the filling of new *5g* and *6f* inner subshells. Electron shell *8* is also new. However, to our present knowledge, it is unlikely that elements beyond 126 will exist. The figure was taken from *Scientific American*, April 1969.

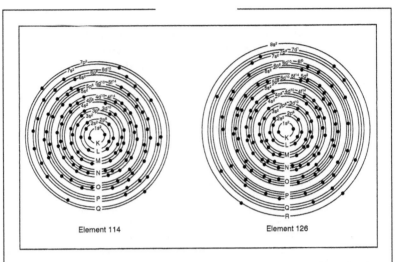

Element 114   Element 126

Figure 1.3 Schematic drawing of the electron-shell structures predicted for the
superheavy elements 114 and 126. In X-ray terminology, the shells
are designated *K* through to *R*; in spectrographic terminology they
are *1* through to *8*. The spectrographic subshells are *s, p, d, f* and, in
the case of element 126, also *g*. The maximum number of electrons
(dots) in any *s* subshell is two, in any *p* subshell six, in any *d* subshell 10, in any *f* subshell 14 and in any *g* subshell 18. However, we
should keep in mind that the drawing is schematic. The increasing
radius of the circles indicates the subsequent filling of the orbits but
does not correspond to the real location of the electrons' orbit. Electrons do not move on precisely located circles or ellipses like stars
around the Sun. Such a picture is correct only within a certain
approximation, although it is used very often for illustration purposes. Because *s* electrons, for instance, do not have any angular
momentum, they cannot move in circles around the nucleus. We
can imagine the location of one of those *s* electrons as being inside
a cloud with a diffuse surface and with the nucleus at the centre.
There is no way we can determine where exactly inside the cloud
the almost point-like electron is located; but we very exactly know
the contours of the cloud. The extension of the cloud is small for *1s*
electrons and larger for *8s* electrons. The figure was taken from
Scientific American, April 1969.

characteristic of nitrogen, not carbon; in decaying, the carbon atom has been transmuted into nitrogen like this:

$$^{14}_{6}C \rightarrow ^{14}_{7}N + \beta^-$$

Now, back to Rutherford's first experimental nuclear reaction: the projectiles he used, emitted from a radioactive source, were α-particles, already identified as the nuclei of atoms of the element helium (He). The reaction took place in a glass tube filled with gas. When the gas was nitrogen, he could observe scintillation even at a distance of several tens of centimetres from the source even though the alpha particles had long come to a halt. Rutherford concluded that the scintillation was due to protons ejected with great speed from the nitrogen nucleus by the impact of an alpha particle. The residual nucleus, at the time an unknown isotope of oxygen with mass number 17, was observed and characterised in 1925 with the help of a cloud chamber (developed by Charles Thomson Rees Wilson in 1912 in Cambridge). The chamber was filled with nitrogen gas, the atoms of which are ionised by the passage of charged particles. When the ionisation occurs, molecules of the also present alcohol vapour condense and small drops are formed, leaving a visible trail. Each reaction was observed by following the traces of an ingoing α-particle and an outgoing proton plus heavy recoil nucleus. A photo of the reaction is shown in Figure 1.4. In modern notation, the reaction is written like this:

$$^{4}_{2}\alpha + ^{14}_{7}N \rightarrow ^{17}_{8}O + ^{1}_{1}p$$

The numbers at the corners of each symbol mean the following:

- top left shows the mass number (A), the sum of protons plus neutrons, in which a proton or neutron weighs 1 unit;
- bottom left is the number of protons in the particle and hence the number of positive electric charges it bears; this number is often omitted because it is obvious from the symbol of the element;
- bottom right (when present) signifies how many of those particles are present in the chemical structure (thus carbon dioxide is CO_2, meaning one carbon combined with two oxygen atoms).

In the above equation, 'α' means an α-particle, 'N' is a nitrogen atom, 'O' is an oxygen atom and 'p' is a proton.

Note that in every nuclear reaction the sum of the number of protons and the number of neutrons is the same on each side of the equation.

In 1919, the neutron was not yet known; it was discovered in 1932 by James Chadwick in Cambridge using the reaction

$$^{4}_{2}\alpha + ^{9}_{4}Be \rightarrow ^{12}_{6}C + ^{1}_{0}n$$
(Be = beryllium; n = neutron)

Figure 1.4 The nuclear reaction $^4\text{He} + {^{14}\text{N}} \rightarrow {^{17}\text{O}} + {^1\text{H}}$ made visible in a cloud chamber. This was the first nuclear reaction discovered by Rutherford. The photo (which was taken later) shows the traces of α-particles (He) emitted from a source at the bottom inside the cloud chamber. The chamber was filled with nitrogen gas as target (invisible). The proton (H) is emitted from the location of the reaction to the lower right (long thin line), the ^{17}O recoil nucleus to the upper left (short thick line). The figure was taken from Blackett and Lee (1932).

The discovery of the neutron solved many problems about the composition of the nucleus and enabled Werner Karl Heisenberg in Leipzig and independently Dmitri Dmitrievich Ivanenko in Moscow to work out that it is composed of both protons and neutrons.

The energy of the α-particles emitted in radioactive decay, and later exact mass measurements of isotopes, showed that another force (the *nuclear force*), much stronger than the known forces of gravitation and electromagnetism, must be involved in the processes. This nuclear force is the third force acting between our bricks but only between the *nucleons*, the comprehensive term for protons and neutrons. The range of the nuclear force is short, not more than the diameter of a nucleon. This is very different from the long range gravitational and electromagnetic forces which both diminish to infinity in proportion to distance (according to the *inverse square law*: the force decreases in proportion to the square of the distance – twice as far, one quarter the strength; three times as far, one-ninth of the strength and so on). The nuclear force is the strongest of the four known forces.

The early 1930s also saw George Gamow (in Leningrad), Hans Albrecht Bethe (Munich) and Carl Friedrich von Weizsäcker (Leipzig) develop the first model for the nucleus, one which quantitatively described many nuclear properties quite well, especially nuclear binding energy. It was called the *charged liquid drop model,* reflecting the basic physical properties underlying its structure.

Just as molecular forces acting on its surface keep a drop of water together, it was assumed that the nucleons form a tiny drop of nuclear matter. The weight and the volume of the drop increase with the number of protons and neutrons. Due to surface tension, the nuclear drop has a spherical shape. Even a very special property, the binding energy which is related to the strong nuclear forces, is well described by the model. When the nucleons join together to form a drop, the nuclear forces start to act. Because the nucleus is more stable than the separated nucleons a certain amount of binding energy is set free during the formation process. This is similar to what happens in chemical reactions when the reactants become hot. In the case of nuclear matter, the 'heat' is emitted as high energetic electromagnetic radiation, the so-called gamma rays, abbreviated as the Greek letter γ. As a consequence, the weight of the nucleus formed is slightly less than the sum of the weights of the constituents, the protons and neutrons. This is possible because mass can be transformed into energy, a phenomenon which is described well by Einstein's law $E = mc^2$.

The theoretical background for an understanding of the static and dynamic properties of elementary particles was provided by the newly developed *quantum theory.* The main progress in this new exciting field of theoretical physics was made in the 1920s by theoreticians like Max Born (Göttingen), Paul Dirac (Cambridge), Werner Heisenberg (Munich, Göttingen, Leipzig), Wolfgang Pauli (Zürich), Erwin Schrödinger (Zürich, Berlin, Graz) and others. Many of the properties of elementary particles which could then successfully be explained, or were predicted, carry the name of the developer: the *Born approximation, Dirac electron theory, Heisenberg uncertainty principle, Pauli law,* and the *Schrödinger equation,* just to mention a few.

In early radioactivity studies there was a lot of confusion about the origin of the electrons observed in nuclear transformations. A clear distinction between electrons emitted from the nucleus – these were the β-rays – and those emitted from the cloud of electrons surrounding the nucleus was made in 1919 when Chadwick showed that the β-particles had a continuous energy spectrum (that is, they were emitted with varying speeds) while the atomic electrons were monoenergetic (they all travelled at the same speed). However, the continuous energy spectrum seemed to violate the law of energy conservation and, as early as 1930, this led Pauli to postulate a new particle, the neutrino, and Enrico Fermi in Rome to develop a whole theory of nuclear β-decay in the following years.

The basic process can be written as a transmutation of the neutron:

$$n \rightarrow p + \beta^- + \bar{\nu}$$

In this process the neutron decays into a proton (p), positively charged, an electron (β^-), negatively charged, and a neutral antineutrino $\bar{\nu}$ (pronounced 'nu-bar'). One heavy particle, the neutron, decays into three, a similar heavy one, the proton, and two light ones. This is possible because the neutron is slightly heavier than the proton. The two light particles belong to the particle family which is called the *leptons* (meaning light particles) whereas protons and neutrons are *baryons* (heavy particles).

It was observed very early in particle physics that particles within a family are created in pairs or that certain pairs mutually annihilate one another into electromagnetic energy, the gamma rays. This behaviour suggested the concept of *particles* and *antiparticles*. When particles are created in a decay process or in a high energetic collision, they are always created in pairs, as a particle and its antiparticle. If the particle bears an electric charge, then the antiparticle bears the opposite charge; the net charge is always zero. If we count the antiparticle as a negative of the particle, then the net number of particles is also zero. We may expect therefore that somewhere in the universe there exists another world consisting of antiparticles. Physicists are still searching for this *antimatter*. Alternatively, the law may have been violated during the creation process of the world and more matter than antimatter was produced. Physicists are also looking at this possibility.

In β^- decay one baryon, the neutron, changes into another baryon, the proton. In order to balance the charge, a lepton is created which carries negative charge, the β^-. However, this β^- cannot be created without an antiparticle of the lepton family, this is the antineutrino $\bar{\nu}$.

As we saw above in the case of ^{14}C, if β^- decay involves a neutron inside a nucleus, the resulting proton remains within the nucleus which increases the net charge by +1 and changes the element into the next one up in the Periodic Table. Using this process, Fermi and co-workers in Rome tried in 1934 to produce new elements beyond uranium, which, at number 92 was then the heaviest element known.

For any particular element, β^- decay is energetically possible only for isotopes having more neutrons than the stable form. In a graph of protons (Z) versus neutrons (N) (Figure 1.5), those nuclei, coloured blue, are located below and to the right of the stable nuclei (in black), and quite distinct from them. Nuclei above and to the left of the stability line are marked in red; they have a relatively high proton-to-neutron ratio and can decay back to the stability line by emission of a positron (an 'anti-electron') and a neutrino. The basic process is:

$$p \rightarrow n + \beta^+ + \nu$$

This transmutation can take place only with protons present in a proton-rich

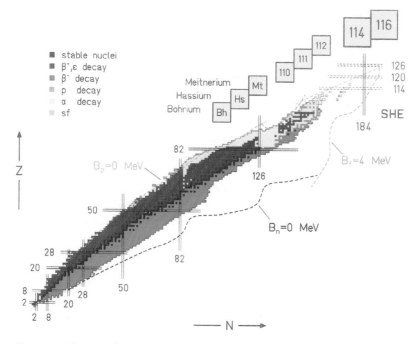

Figure 1.5 The chart shows all the known nuclei in a coordinate system of their proton number (Z) and neutron number (N). The 'magic' numbers are indicated by vertical and horizontal double lines. The double dashed lines mark the uncertainty for the location of superheavy elements at proton number 114, 120 or 126. Black squares are nuclei which occur in nature, almost all of them are stable nuclei. Radioactive nuclei are coloured according to their decay mode: β-minus decay in blue, β-plus decay in red, α-decay in yellow and spontaneous fission (sf) in green. The dashed single lines indicate the theoretically predicted limits of the chart. Few nuclei beyond the proton-drip line ($B_p = 0\,\text{MeV}$ means zero proton binding energy) of elements between $Z = 50$ and 84 decay by proton emission (orange). The neutron-drip line ($B_n = 0\,\text{MeV}$ means zero neutron binding energy) has been reached only for light elements. A small fission barrier ($B_f = 4\,\text{MeV}$) limits the existence of nuclei in the region of heavy elements. The six elements, bohrium to 112, discovered in Darmstadt and 114 and 116 presently investigated in Dubna are framed. The claim for observation of element 118 in Berkeley was retracted by the authors in July 2001, two years after the announcement.

nucleus, not to a free proton (luckily, as it turns out, otherwise we would not exist!). The element number is reduced by one unit. An example is the decay of yet another carbon isotope, ^{11}C, the nucleus of which has its normal six protons but only five neutrons; by the emission of a positron, one proton is turned into a neutron so the resulting atom is beryllium with five protons and six neutrons:

$$^{11}_{6}C \rightarrow ^{11}_{5}Be + \beta^+ + \nu$$

The positron, β^+, is actually an anti-electron, an anti-particle in the lepton family and the neutrino, ν, is the particle. Again, the net number of particles and anti-particles is conserved.

Positron emission was first observed by Irène and Frédérique Joliot-Curie in 1934 in Paris. For the sake of completeness (and with apologies for the added complexity), we do just need to mention that, instead of emitting a positron, a proton inside the nucleus can also capture an electron from the outer electron cloud and emit a neutrino. This is *electron capture*, abbreviated to 'EC' or 'ε' (epsilon), but the result is similar to β^+ emission. Electron capture dominates β^+ emission in the region of heavy elements. This is easy to understand. Nuclei of heavy elements attract the inner electrons of the surrounding electron cloud so strongly because of their high charge so that a relatively large part of the orbit is already inside the larger nucleus. In this case it is easier for the nucleus to capture such an electron instead of emitting a positron but in both cases the neutrino is emitted.

Compared with α-decay and other forms of radioactivity, β-decay is relatively slow; half-lives for β decay are never shorter then a few milliseconds, but also seldom longer than a year. This is because the force acting between leptons and baryons is weak – that is why this fourth force acting between subatomic particles is actually called the *weak force*. Its range is very short, about 100 times shorter than the nuclear force; its interaction therefore takes place inside the nucleon, proton or neutron.

We have nearly reached the end of our preliminary exploration. Only two more discoveries, both of them contributing substantially to the atomic scene, remain briefly to be mentioned.

The first was the *fission process*, discovered by chance at the end of 1938 by Otto Hahn and Friedrich Wilhelm (Fritz) Straßmann in Berlin-Dahlem: heavy nuclei can break in two, releasing a huge amount of nuclear energy – about 20 times more than in α-decay. The phenomenon they observed was explained by Lise Meitner (who had to leave the group and go to Sweden in summer of that year because of political developments in Germany). Fission could readily be understood on the basis of the nuclear liquid drop model and it indicated a possible end of the Periodic Table.

Box 1.2 The concept of 'half-life' and 'lifetime'

The decay of radioactive isotopes is a statistical phenomenon. One cannot predict when any particular atom, or group of atoms, will emit a particle, only that *in a population of a certain radioisotope*, half the atoms will decay within a constant period. Let's take an example, that of $^{14}_{6}C$. This is the isotope used for radio chronometry, the method used, for example, in determining that the age of the ice man called 'Oetzi', who was found in 1991 in the Alps near the Austrian–Italian border, was 5,300 years. Suppose we watched a collection of 1,024 atoms of this material, we would find that over the course of 5,730 years about half of them (512) would decay. If we continued observing for another 5,730 years, half the remaining population (256) would decay … and in the next 5,730 years half of *that* remaining population (128) would blow up, and so on. With a large starting population you are never in theory left with none; you simply go on approaching zero, halving the remaining number of atoms of that type in the population each period.

Half-lives are characteristic for each isotope and no ways have been found to influence them. For $^{14}_{6}C$, as we have seen, the half-life is 5,730 years; for a second radioisotope of carbon ($^{11}_{6}C$) it is only 20 minutes and for yet another ($^{15}_{6}C$) it is only 2.5 seconds. But two isotopes of this element ($^{12}_{6}C$ and $^{13}_{6}C$) do not decay at all; they are stable and can last forever. One could say that their half-lives are infinitely long.

How do we measure half-lives? This is easy if the half-life is not unusually short or long and if there is a large quantity of the radioactive isotope available. From each decay we measure the emitted particle with a detector. Then we count the number of decays for a while, let say for 10 minutes. We come back after a year and count the same sample again. If the number counted is now half of that of the first measurement, then the half-life is one year.

But what can we do if we produce only one atom at a time? In this case we must know very precisely the time when it was produced as well as the time when it decayed. The period between is the lifetime. There exists a well-defined mathematical relation between half-life and lifetime: the half-life is 30% shorter than the lifetime. To be precise, the difference is given by the natural logarithm 2 which has the value 0.69 . . .

If a species of nuclei has a lifetime of two seconds for example, it does not mean that each of these nuclei will decay after two seconds. Some will already have decayed after one second, some after five seconds and a few after 10 seconds or even later. From measurements we get the mean value for the lifetime as an average of all the individual data. The more data we measure, the smaller is the error bar for this mean value. The laws of statistics tell us that the error bars for mean value obtained from 100 events are ±10%, from 10,000 events they are ±1% and so on. Even if we measured very few events we still get error bars which are large and asymmetric, in the extreme case of only one event measured we obtain error bars of +480% and −46%. For two events the errors are already considerably reduced +180% and −39%.

Hahn had been studying radioactivity since 1904. He worked together with Sir William Ramsay in London and with Rutherford in Montreal. He had discovered a number of new isotopes and new elements: radiothorium ($Z = 90$) in 1904–5, radioactinium (93, 1905–6) and, with Meitner, mesothorium (now radium) (88, 1907) and protactinium (91, 1917). In 1921 Hahn described isomerism and the first isomeric state of a nucleus in ^{234}Pa. Isomerism is spread across the entire chart of nuclei and also plays an important role in the region of the heaviest elements. Nuclei are described as 'excited' when one or more nucleons do not occupy the energetically lowest possible orbits. Usually, these nucleons move back into the orbits which form the *ground-state* of a nucleus within an extremely short time. Gamma rays are emitted to balance the energy. Sometimes, however, the properties of the orbits involved are such that the emission of gamma rays is hindered. Then the lifetime of the excited state gets long, milliseconds, seconds, or even years, the excited state becomes isomeric. The chemical behaviour of nuclei in an isomeric state is identical to that of the element to which they belong.

The fission process discovered by Hahn and Straßmann in 1938 was induced by irradiation of uranium with neutrons ('neutron induced fission'). In this case, the fissioning uranium is in an excited state. Two years later, Georgi Nikolaevich Flerov and Konstantin Antonovitsch Petrjak, working at the Radium Institute in Leningrad, observed that uranium also decays by fission when it is in the ground-state and without being irradiated by neutrons. This decay mode is called 'spontaneous fission'.

The second discovery was more theoretical. It was known from earlier experiments that nuclei with certain numbers for the protons and neutrons exhibit special properties: they are spherical and their nuclear binding energy is exceptionally high. These 'magic' numbers were 2, 8, 20, 28, 50, and 82 and, for the neutrons, 126 as well (Figure 1.5). In 1949, the nuclear shell model was introduced by Maria Göppert-Mayer working at Chicago and also independently by Otto Haxel, J. Hans Daniel Jensen and Hans Eduard Suess in Heidelberg. The Bohr atomic model which so excellently explained the arrangement of electrons in shells surrounding the nucleus was not applicable for the nucleus itself. The difference is that the electromagnetic forces have a long range and that the tiny heavy nucleus forms a centre for the orbiting electrons, whereas the nuclear forces have short range and there is no centre around which the nucleons can move. It was to the credit of Göppert-Mayer, Haxel, Jensen and Suess that they introduced a strong coupling between the motion of orbiting nucleons and their own spin. The spin of a particle is the rotation around its own axis. The so-called spin-orbit force resulted in an arrangement of the orbits so that one shell was filled with a number of protons or neutrons which was in agreement with the empirically known magic numbers. The surface tension took over the task of keeping the nucleons together and giving a shape to the nucleus.

The most stable nuclei are those in which the number both of protons and neutrons is 'magic'. The heaviest known 'doubly magic' nucleus is the most abundant isotope of lead in which $Z = 82$ and $N = 126$. Whereas the liquid drop model suggested an end of the Periodic Table at around element 110 due to fission tendencies, the shell model offered hope that a next doubly magic nucleus beyond ^{208}Pb could again exhibit increased stability. The idea of superheavy elements (SHEs) was born and a rush to find them began.

2

Elements created in stars

Perhaps 15, 18 or 20 billion years ago, our universe started with a 'Big Bang'. Details are very uncertain and, anyway, nobody knows the details for sure. But, however it started, the universe has been expanding for a very long time, initially an enlarging cloud of gas consisting of elementary particles cooling down as expansion proceeded.

Most of the material then in existence was a mix of protons, electrons and neutrinos (all three stable as far as we know) and, of course, electromagnetic radiation. Free neutrons disappeared because, within minutes of formation, they decayed into protons, electrons and antineutrinos. A neutron could survive only when it was caught by a proton to form a deuteron nucleus which is also stable. (A deuteron is a form of hydrogen nucleus containing one proton – like hydrogen – plus one neutron. It is therefore chemically just like hydrogen but, of course, twice as heavy and that gives it some slightly different properties; indeed, it is called 'heavy hydrogen' or 'deuterium', abbreviated as 'd', 'D', or '^{2}H'. About 0.015% of natural hydrogen is in this form.) Two deuterons could form a stable helium nucleus. When the temperature of the expanding gas was still high, relatively small amounts of lithium ($Z = 3$) and beryllium ($Z = 4$) were also created.

The early universe was by no means homogeneous but rather 'lumpy' in its distribution of matter. Some collections or clouds of particles developed as the result of gravitational attraction. About 12 billion years ago, in regions of the highest density where the temperature also was at its highest, the light elements began to 'burn' in the first stars and galaxies. (We put 'burn' in quotation marks because it is really a slang term for high-temperature fusion, not the sort of burning we see in a terrestrial fire. We use similar terminology to describe the rate at which a new company uses up ('burns') its money; of course, the company does not actually set fire to its cash (although it might sometimes seem like that to the investors!). Stars have various masses and hence various interior temperatures, ranging from 10 million to several 100 million degrees, enough to cause hydrogen atoms to fuse into helium. In 1939, Hans Bethe, at that time at Cornell

University in Ithaca, suggested a mechanism for the hydrogen-to-helium fusion:

$$^1_1p + {}^1_1p \rightarrow {}^2_1D + e^+ + \nu + 1.4\,\text{MeV}$$

(two protons combine to form one deuteron, one positron and one neutrino – and release energy. The energy unit in nuclear physics is MeV, million electron volts. This is the energy that a particle carrying one unit of charge gains when it is accelerated by an electric voltage of 1 million Volts. Often used multiples and submultiples of units, their meaning and abbreviation, are given in Table 2.1):

$$^2_1D + {}^1_1p \rightarrow {}^3_2He + \gamma + 5.5\,\text{MeV}$$

(a deuteron fuses with another proton to form light helium plus more energy);

$$^3_2He + {}^3_2He \rightarrow {}^4_2He + 2{}^1_1p + 12.9\,\text{MeV}$$

(two light helium atoms fuse to form one heavy helium and two protons; yet more energy is released).

It is worth noting, by the way, that the energy resulting from such fusion reactions of the lightest elements is only about 10% that of the fission of an uranium atom. However, there is much more hydrogen on earth than uranium, a circumstance one has to consider when thinking about long range plans to solve the energy problem. All the same, the energy released

Table 2.1 Prefix names of multiples and submultiples of units

Factor by which unit is multiplied	Prefix	Symbol
10^{18}	Exa	E
10^{15}	Peta	P
10^{12}	Tera	T
10^{9}	Giga	G
10^{6}	Mega	M
10^{3}	kilo	k
10^{-3}	milli	m
10^{-6}	micro	μ
10^{-9}	nano	n
10^{-12}	pico	p
10^{-15}	femto	f
10^{-18}	atto	a

maintains the products in a state of very rapid motion, preventing the stars collapsing under their own gravitational attraction and keeping them shining.

In large (and therefore hotter) stars, three helium atoms can fuse into one carbon, a process which also goes on in dying, collapsing stars when most of their hydrogen has already been converted into helium and there is insufficient kinetic energy generated to prevent collapse. As the star contracts, its interior temperature rises and ^{4}He is fused into ^{12}C. The resulting radiation pushes outwards the outer layers of the star which consequently cool down, thus creating a 'Red Giant'.

At high temperatures, helium can also fuse with carbon to form oxygen. Under strong gravitational contraction, at temperatures ranging up to one billion degrees, such fusion processes continue up the Periodic Table as far as iron and nickel; for heavier elements, however, energy is no longer set free by fusion but energy input becomes a requirement and the reaction changes from being 'exothermic' to 'endothermic' ('energy out' to 'energy in'). By using a telescope to look at the spectral lines in its light, it is possible to explore the distribution of the elements in various stages of a star's lifetime and as a function of its mass. This can be done because light emitted from hot gases does not consist only of a continuous distribution of light rays of all kind of colours (which means of all energies), but discrete, mono-energetic rays are also emitted. These are attributed to certain transitions of electrons from one orbit to another stronger bound orbit, and the energy or colour is characteristic for a particular element. The spectral lines become observable when light is analysed with a prism. That is how, in 1868 during the solar eclipse, Pierre Jules César Janssen at the observatory in Meudon south-west of Paris obtained evidence for the existence of the element helium, long before it was known on Earth. Not until 1895 was it discovered by Sir William Ramsay working on uranium minerals at the University of London; the helium he had found in his experiments was generated in the form of α-particles originating from the decay of radioactive atoms in the ores.

How (and where) were the heavier elements produced in nature?
Two possibilities exist which differ according to the time scale of the process, both based on neutron capture and subsequent β$^-$-decay.

At low neutron densities, the slow 's-process' takes place over thousands of years; it is believed to occur in the red giant phase of stellar evolution. Neutrons are added to the existing elements up to iron and nickel which had earlier been produced by fusion. The neutron capture rate is low so that there is time for the isotopes produced to undergo β$^-$-decay to the next heavier element, which again captures a neutron. Step by step, the element production ladder is climbed. This path follows essentially the β-stability line (Figure 2.1). This process stops with polonium (Z = 84) and astatine (Z = 85) for which the α-half-lives, close to the β-stability line, are so short that they rapidly decay into lead (Z = 82) and bismuth (Z = 83), respectively.

Nevertheless, this s-process does produce target nuclei for the rapid 'r-process' in which neutron additions take place at neutron densities of 10^{20}–10^{25}/cm^3 with reaction times in the order of 1–100 seconds (10^{20} means 100 billion billions, written out, this is a 1 followed by 20 zeroes). Many tens of neutrons may be added to each target nucleus, so bypassing the α-gap above lead and bismuth on the extremely neutron-rich side, where the isotopes again undergo β-decay, thus increasing the element number of the decaying nucleus.

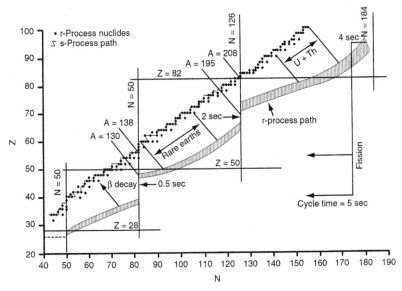

Figure 2.1 Neutron capture paths in astrophysical nucleosynthesis for the (slow) s-process and the (rapid) r-process. The s-process follows a path in the *NZ* plane along the line of beta stability and is equivalent to the neutron capture process occurring in nuclear reactors. It ends at element 83 due to short alpha half-lives of elements 84 and 85. The neutron-rich progenitors to the stable r-process nuclei, which are here shown as small circles, are formed in a band in the neutron-rich area of the *NZ* plane such as the shaded area shown here. So far all nuclei in the shaded area are unknown except a few cases at Z = 28. After the synthesising event, the nuclei in this band β-decay to the stable r-process nuclei. The figure was taken from Crandall (1974).

In nature, high neutron densities are associated with supernovae and quasars. The rapid neutron capture path essentially follows the line of high neutron excess. There, the β^--half-lives are short (<100 milliseconds) and neutrons are virtually not bound to the nucleus – we say their binding energy is small. It may even become negative: neutrons are then no longer bound to the nucleus but are emitted forming lighter, more stable isotopes. The smooth r-process path is broken at the magic neutron numbers 50, 82 and 126 at which shells for the neutrons are filled (see Figure 2.1). An additional neutron (the 51st, 83rd and 127th) is then especially loosely bound to the nucleus and can be easily emitted. This is similar for the electrons of elements just above the noble gases, e.g. sodium above neon. These electrons are only loosely bound and the atoms of those elements tend to form ions by emission of the almost unbound electron outside the closed shell.

Therefore, at the closed neutron shells, an equilibrium develops between neutron capture and the reverse neutron emission process. The result is a lengthened effective neutron absorption life which allows extensive β^--decay while these magic neutron numbers are being traversed. During and after a neutron burst of a supernova explosion, the extremely neutron-rich isotopes decay back in the direction of the β-stability line by converting neutrons into protons thus forming heavier elements.

Heavy nuclei tend to fission, breaking up into two parts. The two fragments, a lighter and a heavier one, are neutron rich and β^--decay into the direction of the stability line. They again serve as fuel for the r-process. In this way the so-called *fission cycle* is created. The probability of fission is increased if the nuclei are not in the ground-state (the lowest possible energy state) but in a state of higher energy (an excited state). Excited states are created after neutron capture. In the heavy elements neutron-induced fission thus competes with neutron capture. Furthermore, excited states of nuclei are also populated after β^--decay. Fission occurs (β-*delayed fission*) which produces at least an intermediate termination of the r-process between mass numbers 258–263 at around element number 100 (fermium, Fm).

How far this β^--delayed fission block extends into the region of superheavy elements is unknown. It does, of course, make all the difference for observing superheavy elements in nature.

The abundance of elements and isotopes in nature is calculated in astrophysical models. Comparison of the results with the measured abundance is a sensitive method to prove the reliability of the models. The calculations are complicated by the fact that the r-process nuclei (those in the shaded area in Figure 2.1) are unknown, except a few very light nuclei. Predicted lifetimes and binding energies from nuclear models must be used, which, however, become more and more uncertain with increasing distance from the known nuclei. Experimental investigation of r-process nuclei is one of the necessary tasks in present day nuclear physics and astrophysics, and it is closely related to the synthesis of superheavy elements in nature.

3

Synthetic elements

Early attempts at synthesising elements – neptunium and plutonium

'The first scientific attempts to prepare the elements beyond uranium were made in 1934 in Rome by Enrico Fermi, Emilio Segrè and their co-workers shortly after the existence of the neutron was discovered' (Seaborg and Loveland 1990). Neutrons were found to attach easily to many elements, producing isotopes of the initial elements which emit β^- rays and hence decay into the element with the next higher atomic number. If that same sort of capture should occur to uranium ($Z = 92$) one would expect the formation of element 93.

At that time, neutrons were produced via a nuclear reaction between α-particles (from radium [$Z = 88$] or a similar source) and beryllium ($Z = 3$):

$$^4\alpha + {}^9Be \rightarrow {}^{12}C + n$$

Stable beryllium is monoisotopic, i.e. it consists only of the isotope ^9Be. In the experiment, the beryllium powder was well mixed with the radioactive α-emitter. The neutrons were emitted at various energies (i.e. at various speeds); the highest probability of capture by uranium nuclei, however, is for low speed neutrons, moving at a speed in thermal equilibrium with the surrounding material (paraffin). The thermalisation process is fastest when the neutrons are scattered at the other light nuclei present in paraffin. In each scattering process the neutrons lose speed (energy) until they have the same speed (i.e. the same energy or temperature) as the nuclei in the surrounding paraffin. Such a paraffin moderator, as it was used by Hahn, Meitner and Straßmann, is shown in Figure 3.1.

The team in Rome found a number of radioactive products which they thought might be new elements. Their idea was that, if element 93 would indeed emit β^--rays, element 94 must be produced. Indeed, at least five β-ray emitters were observed which differed by their half-lives of 10 seconds, 40 seconds, 13 minutes, 90 minutes and about one day. However, chemical studies by Hahn, Meitner and Straßmann – later continued in December 1938 without Lise Meitner, as mentioned earlier – showed that these species were isotopes of the known lighter elements formed by the fission of uranium.

Figure 3.1 Neutron sources (left) and paraffin moderator used by Otto Hahn in experiments for the synthesis of transuranium elements which resulted in the discovery of fission. The figure was taken from *History of the discovery of nuclear fission*, Technical University Berlin, 1989.

Element 93 was found at the University of California at Berkeley in spring 1939, in an investigation into the energy of the two fission fragments from neutron-induced fission. Edwin McMillan had placed a thin layer of uranium oxide on a piece of paper. Next to it he

> stacked very thin sheets of cigarette paper to stop and collect the uranium fission fragments. In the course of his studies he found there was another radioactive product of the reaction, one that did not recoil sufficiently to escape the uranium layer as did the fission products. He suspected that this product was formed by the capture of a neutron by the more abundant isotope of uranium, ^{238}U. In 1940 he and Philip H. Abelson, who joined him in this research, were able to show by chemical means that the product was an isotope of element 93 . . .
>
> (Seaborg and Loveland 1990)

Element 93, called *neptunium*, ^{239}Np, is formed in the way anticipated by Fermi and his colleagues. Surprisingly, the chemical properties of neptunium were similar to those of uranium and not like those of rhenium as suggested by the Periodic Table of that time. 'Neptunium, the element beyond uranium, was named after the planet Neptune just because this planet is beyond the planet Uranus for which uranium itself had been named' (ibid.).

Immediately thereafter, during the summer and autumn of 1940, McMillan started looking for the β^--decay daughter product of ^{239}Np which obviously would be the isotope with mass $A = 239$ of element 94. But he found nothing. The reason (as we now know) was the long half-life of ^{239}Pu at 24,000 years, giving such low rates of radiation from his sample that he had difficulty detecting it. He began to bombard uranium with deuterons in the 60-inch cyclotron in the hope he might find a shorter-lived isotope – one with a higher intensity of radioactivity which would be easier to identify and measure. Before he could finish this project, war was declared and he was called away to work on radar at the Massachusetts Institute of Technology.

His work was continued by Glenn Seaborg together with Joe Kennedy and his student, Arthur C. Wahl. On December 14, 1940, they performed the first deuteron bombardment of uranium, from which Wahl isolated a chemical fraction containing element 93. They observed that the radioactivity of this fraction had characteristics different from the radiation from a sample of the known isotope ^{239}Np; the most significant difference was that this sample exhibited α-radioactivity. They concluded that the α-activity was due to a daughter of the new neptunium isotope (^{238}Np) with a half-life of approximately 50 years (most recently valued at 87.74 years). In that daughter product they had identified for the first time an isotope of element 94.

The reaction and decay chain for the synthesis of ^{238}94 were:

$$^{238}U + {}^{2}H \rightarrow {}^{238}Np + 2n$$

(uranium-238 plus a deuteron gave neptunium 238 plus two neutrons)

$$^{238}Np \rightarrow {}^{238}94 + \beta^- + \bar{\nu}$$

(neptunium-238 decayed to plutonium-238 with the emission of a β-particle and an antineutrino).

In March 1942, element 94 was christened 'plutonium' with the chemical symbol 'Pu', 'named after the planet Pluto following the pattern used in naming neptunium. Pluto is the second, and last, known planet beyond Uranus' (Seaborg and Loveland 1990).

The man-made s-path – americium, the element from a nuclear reactor

All large-scale transuranium production was and is based on capture reactions with neutrons. The largest terrestrial target for neutrons is the fuel in a nuclear power reactor. However, only one element, *americium* ($Z = 95$), the fourth transuranic to be discovered, was identified for the first time in a nuclear reactor. The isotope ^{241}Am was identified by Glenn T. Seaborg, Ralph A. James, Leon O. Morgan and Albert Ghiorso late in 1944 at the wartime

Metallurgical Laboratory (now the Argonne National Laboratory) near Chicago as the result of successive neutron capture reactions by plutonium isotopes in an uranium graphite reactor ('γ' signifies emission of high energy gamma rays):

$$^{239}Pu + n \rightarrow ^{240}Pu + \gamma$$
$$^{240}Pu + n \rightarrow ^{241}Pu + \gamma$$
$$^{241}Pu \rightarrow ^{241}Am + \beta^- + \bar{\nu}$$
$$^{241}Am + n \rightarrow ^{242}Am + \gamma$$
$$^{242}Am \rightarrow ^{242}Cm + \beta^- + \bar{\nu}$$

Actually the element *curium* (Cm) had been produced and identified a few months earlier after irradiation of plutonium with helium ions. The problem in separating americium and curium after the reactor irradiation was their similar chemical behaviour which led Seaborg to establish the concept of the actinide series in 1944 (see Box 1.1 and Figures 1.1 and 1.2). He thought that perhaps all known elements heavier than actinium were misplaced in the Periodic Table. The theory advanced was that these elements might constitute a second series similar to the series of rare earth or lanthanide elements. Seaborg was awarded the Noble Price of Chemistry for this work in 1951.

'The key to their chemical separation, later achieved at Berkeley, was ion-exchange' (Seaborg and Loveland 1990). This is the same method used domestically for water softening. One type of ion in a solution (e.g. calcium in hard water, which causes lime deposits) is bound to a mineral (usually natural zeolite) and a less troublesome ion (e.g. sodium) is released into the solution. Such ion exchange reactions are widely used for purification, concentration and ion separation as well as for the separation and identification of transuranium elements. The actinide series ends with element 103, *lawrencium* (Lr); element 104, *rutherfordium* (Rf) clearly belongs to the IVth group of the Periodic Table with chemical properties resembling titanium, zirconium, and hafnium (see Figure 1.1).

In terms of electron configurations, an inner shell (numbered 4f) is filled successively for the lanthanides. The shell becomes full with 14 electrons which results in a total of 15 lanthanide elements including lanthanum itself, the element from which the series was named and which has an empty 4f electron shell. (For details of the nomenclature see Box 1.1 and Figures therein). Similarly, for the actinides the 5f shell is successively filled, again with 14 electrons resulting in 15 actinide elements. The consequence of such internal filling is that the outer electrons remain much the same. It is these outer electrons which primarily determine the chemical behaviour, similar for all the elements in each of the lanthanide and actinide series.

The analogy of the chemical properties of elements 95 and 96 with the corresponding elements of the lanthanide series was also the reason for their names. '"Americium" was suggested for element 95 by analogy with the naming

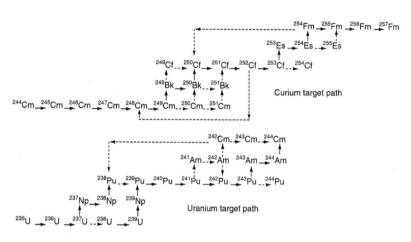

Figure 3.2 Neutron capture paths used to synthesise transuranium nuclei in nuclear reactors. The horizontal arrows pointing right represent neutron capture, vertical arrows pointing up represent β^--decay and vertical downwards arrows are electron capture decay. The sequence of long dashed arrows represents α decay. The main paths are marked by solid arrows, whereas weaker branches are indicated by dashed arrows. The figure was taken from Crandall (1974).

of its rare earth counterpart (*europium*) after Europe' (Seaborg and Loveland 1990). The first americium isotope to be discovered, ^{241}Am with a half-life is 432 years, is now well-known as one of the α-energy calibration standards.

The large-scale production of transuranium elements in nuclear reactors is based on two factors:

- the large number of neutrons and target nuclei available; and
- the lack of electrical charge of neutrons which allows them to travel long distances through the target material and to react with other nuclei without being electrically repelled.

How isotopes and elements are produced by neutron capture in nuclear reactors can be best explained using a two-dimensional map of nuclei spanned by mass number and element number as co-ordinates. Such a plot is shown in Figure 3.2. The arrows mark the path for production of nuclei by successive neutron capture, β^-, β^+ and α decay. Note that nuclides used as reactor fuel (^{235}U, ^{238}U, ^{239}Pu) also lie on the same path.

Starting with ^{238}U, the main path goes to ^{239}Pu which undergoes multiple neutron capture to ^{243}Pu; the latter decays to ^{243}Am which, in turn, undergoes capture and decays to ^{244}Cm. ^{244}Cm can undergo multiple neutron captures to yield the Cm isotopes to ^{249}Cm, while subsequent captures and decays lead to ^{249}Bk (berkelium), ^{250}Bk, ^{250}Cf (californium), ^{251}Cf, ^{252}Cf and so on up to ^{257}Fm (fermium), where the chain ends due its very rapid spontaneous fission of ^{258}Fm.

(Seaborg and Loveland 1990)

This information is so appealing to me because, when reading it again a little bit faster than usual, I feel almost like a neutron producing elements myself! From the same book and from other publications of Ghiorso and Seaborg I obtained the intimate knowledge about the early heavy element research in Berkeley described in this and the following chapter.)

In power reactors, however, no appreciable concentrations of nuclides above ^{244}Cm are found. To produce these heavier nuclides requires special reactor irradiations of separated nuclides like the High Flux Isotope Reactor at Oak Ridge in Tennessee. Irradiation of a ^{252}Cf target, which has a half-life of 2.6 years, yields einsteinium and fermium, mainly as ^{254}Es and ^{257}Fm with half-lives of 276 and 100 days, respectively. The world stock of these isotopes is a few micrograms.

The man-made r-path: einsteinium and fermium, elements from atomic explosions

'The most efficient way of producing the transuranium nuclei is via a nuclear explosion' (ibid.). These led, in the early 1950s, to the discovery of the elements 99 and 100 which were discovered unexpectedly in debris from the thermonuclear explosion 'Mike' in the Pacific on November 1, 1952. 'Debris from the explosion was collected, first on filter papers attached to airlines which flew through the clouds. Later, in more substantial quantity, fall-out material was gathered from the surface of a neighbouring atoll' (ibid.) and taken to a number of American laboratories for chemical investigations of the explosion.

> Early analysis, at the Metallurgical Laboratory (Argonne) near Chicago and the Los Alamos Scientific Laboratory in New Mexico, showed the unexpected presence of new isotopes of plutonium, ^{244}Pu and ^{246}Pu; at the time, the heaviest known isotope of plutonium was ^{243}Pu. Sherman Fried, one of the participants, wrote that 'Once having shown it was plutonium-246, it began to dawn on us that, my God, uranium has captured eight neutrons and we still, three weeks later, have enough stuff to see plutonium-246 easily. The way was open'. This observation led to the conclusion that the ^{238}U in the bomb had been subjected to an enormous neutron flux; later calculations showed that an integrated neutron flux of $1-4 \times 10^{24}$ neutrons were delivered in a few nanoseconds, equivalent to a few grams of neutrons!

(ibid.)

Albert Ghiorso described what happened when the news reached Berkeley:

> By the next morning, however, I had come up with a really wild idea. First, I assumed that the curve of the yield at each mass number could be represented by a straight line on a semi-log plot. To get the slope of the line, I assumed that the starting material was ^{238}U and that the relative yield of ^{244}U [which of course would β^--decay to ^{244}Pu] was 10^{-3} [at the time details of 'Mike' were secret]. I felt that this number was the lowest yield that could be seen in the plutonium fraction with certainty on the mass spectrographs of that time. Extrapolating this line to the mass ^{254}U (ten neutrons heavier), which I assumed would β^--decay all the way up to element 100, indicated a yield of roughly 10^{-8} or 10^{-9}. For some reason, which I don't recall now, I assumed that the bomb fraction of 10^{14} atoms (of ^{254}U) was obtainable. This meant that we might get approximately a count per minute of α-activity from $^{254}100$ if it had a half-life of about a month. The possibility was real, although pretty far fetched, and we should do something about it.
>
> (ibid.)

The Ghiorso prediction was remarkably accurate.

> Armed with the knowledge of the multi-neutron capture by ^{238}U, scientists at the University of California immediately began to search in the bomb debris for elements beyond californium ($Z = 98$). Ion-exchange experiments demonstrated the existence of a new element and, within a few weeks, of a second new element. [A source of] α-particles of 6.6 MeV energy with a half-life of 20 days was identified as an isotope of element 99 with the mass number 253 while a 7.1 MeV α-activity with a half-life of 22 hours turned out to be an isotope of element 100 with a mass number 255.
>
> (ibid.)

The first identification of element 100 involved no more than about 200 atoms.

Researchers at the Berkeley, Argonne and Los Alamos laboratories suggested the name 'einsteinium' for element 99 in honour of the great physicist, Albert Einstein; for element 100, the name 'fermium' was proposed in honour of Enrico Fermi who extensively studied neutron capture reactions during the mid-1930s. 'The choice of the name fermium for element 100 has proven to be prescient since it is the last element to be synthesised using neutron capture reactions' (ibid.). Fermi died November 28, 1954; he was living when element 100 was named after him, although the name was not then made public.

> Before this work was declassified, and the original discovery experiments could be announced, isotopes of element 99 and 100 were

produced by other, more conventional methods. Chief among them was that of successive neutron capture as the result of intense neutron irradiation of plutonium in a high-flux materials testing reactor in Idaho.

The difference between this method of production and that of the 'Mike' thermonuclear explosion is one of time as well as of starting material. In a reactor, it is necessary to bombard gram quantities of plutonium for two or three years; short-lived, intermediate isotopes of the various elements thus have an opportunity to decay. The path of element production proceeds up to the valley of β stability, equivalent to the slow s-path of element synthesis in stars. In the thermonuclear device, the larger amounts of uranium were subjected to an extremely high neutron flux for a period of nanoseconds. The subsequent β^--decay of the ultra-heavy isotopes of uranium led to the nuclides found in the debris.

(ibid.)

In this case, the decay matched the natural rapid r-path.

Targets in general and pairing effects

In nuclear physics experiments, *targets* are normally stationary and are bombarded by the moving projectiles because the projectile is usually a lot lighter and so much easier to accelerate.

Depending on their purpose, targets may have various shapes and forms. At one extreme, as we have already seen, a target might be the several kilograms of uranium in a reactor with, as projectiles, the neutrons which originate from the fissioning uranium where they also acquire some kinetic energy. However, the neutrons are best captured by the uranium nuclei when they have low 'thermal' energies (characteristic simply of warm particles rather than those shot out at high speed from disintegrating atoms). Neutrons carry no charge and are therefore not rejected by the electric charge of the uranium target nucleus; we describe neutrons as having 'a high capture cross-section'. Thermalisation is achieved with a moderator, a material of low mass and a low affinity for neutron capture. In collisions between neutrons and the nuclei of the moderator, the neutrons lose speed and stepwise their speed approaches the low value equivalent to the temperature of the moderator.

'Wait a minute,' you might say, 'isn't there a mistake here? On the one hand you tell us that uranium fuses with the neutrons and you say it then can make it fission?' No, there is no mistake. The uranium captures the neutron and then it has a 'choice' of fissioning or of emitting the reaction energy by gamma rays resulting in the uranium isotope with mass one unit higher. Fissioning produces more neutrons but where do they come from? And what speed do they have? To understand this. take a look at the chart of nuclei in Figure 1.5. We see that the stable nuclei do not follow the diagonal in the co-ordinate system of neutrons and protons. Heavy elements carry more neutrons than protons so the line of stable nuclei is bent to the right as the element number increases. Uranium, element 92, is already close to the upper end of the chart; its heaviest isotope, the heaviest of all found in nature, has the mass number 238 and carries 146 neutrons. When this nucleus fissions, it does not break into two identical parts of element 46 (palladium), each with the mass number 119. Instead, it splits into a heavier and lighter fragment – and not always the same fragments; there is a range of possibilities but their sum is always equivalent to ^{238}U. The asymmetric fission is a consequence of the shell structure of the nuclei. The heavier fragment is close to a tin isotope of mass number 132. Tin is element 50 so ^{132}Sn carries 82 neutrons. You remember, both 50 and 82 are magic numbers and the nucleus ^{132}Sn is doubly magic and strongly bound. Nuclei around ^{132}Sn form the heavy fragment in the fission process, the lighter fragment resulting from the difference with ^{238}U.

Because ^{238}U has so many neutrons, 146, the fragments have a surplus of neutrons, which they like to reduce in order to gain stability. During each fission process a number of two or three neutrons is emitted, on the average almost exactly 2.5 neutrons. The neutrons are emitted with high speed, about 5% of the velocity of light. In energy this is 1 MeV. These neutrons are thermalised and then caught by other uranium nuclei which again fission. With each fission more and more neutrons become available; a chain reaction starts.

However, not every captured neutron leads to fission. Whether we get more fission or more fusion depends on the reaction energy which heats up the fused system. The two processes are demonstrated in Figure 3.3, which shows a series of photographs of a vibrating liquid drop of macroscopic dimensions. In one case the drop fissions and in the other the structure keeps together and is forming finally a stable spherical drop. Although the phases of vibration at the beginning appear almost identical, the final state is dramatically different. We see that, in addition to very well describing static properties like binding energy, the liquid drop model of the nucleus also explains dynamic properties like fission and, as we will see later, the fusion of two massive nuclei is equivalent in the model to the amalgamation of two liquid drops into one. But, looking for details, fusion is not the inverse of fission.

In order completely to understand uranium fusion with neutrons and fission we need to mention the 'pairing effect': whenever pairs are formed, additional energy is set free because a pair of nucleons forms a more stable configuration than unpaired individuals. (Sounds almost human!) In atomic and nuclear physics, two particles (two electrons, two protons or two neutrons) form a pair when their spin (the particle's rotation around its own axis) is in opposite directions.

Uranium-238 has 92 protons and 146 neutrons. This is the most stable and abundant isotope. All protons and neutrons are coupled in pairs and make ^{238}U relatively stable. Adding one neutron to ^{238}U raises the temperature (excitation energy) of the new nucleus ^{239}U due to the binding

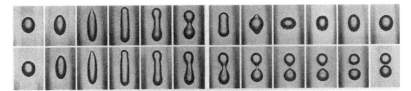

Figure 3.3 Liquid-drop analogy of the phenomenon of nuclear fission is illustrated by these two sequences of photographs selected from a motion-picture film made by S. G. Thompson and his colleagues at the Lawrence Radiation Laboratory of the University of California. The photographs show an ordinary drop of water suspended in oil; the initial deformation of the drop was produced by applying a voltage across the oil. In the sequence at the top, the initial deformation was insufficient to cause fission and the drop has returned to its original, minimum-energy spherical shape. In the sequence at the bottom, the initial deformation and slightly different currents inside the drop has pulled the drop out to its 'threshold' elongation (saddle point), following which it breaks in two at its narrow waist. The two resulting drops quickly round out to the minimum-energy spherical shape, converting the energy of their former deformation into heat. The analogy between the fission of the uranium nucleus and the splitting of a drop of liquid by deformation was suggested by Lise Meitner and Otto Robert Frisch in 1939. The model was developed further by Niels Bohr and John A. Wheeler. The figure was taken from *Scientific American*, August 1965.

energy which is liberated when the neutron is captured. However, when thermalised neutrons are captured the excitation energy is not high enough to fission the resulting ^{239}U. But this is different for ^{235}U, which has 143 neutrons, one of which is therefore unpaired. In that case, the addition of a neutron liberates the binding energy as well as the pairing energy and the resulting more excited or heated ^{236}U will most likely fission.

The pairing effects result in a number of phenomena which depend on the number of particles involved and whether this number is odd or even. For instance, even-Z elements are more stable than odd-Z elements. As a consequence, more stable isotopes exist from even-Z elements. In the case of tin (Z = 50) a total of 10 stable isotopes exist which are all found in natural tin. Odd-Z elements have only one or two stable isotopes, and two elements in the range up to bismuth have no stable isotope at all. These are the artificially synthesised elements, technetium (Z = 43) and promethium (Z = 61). In the region beyond bismuth, only two elements have a half-life long enough, so that they still exist on earth. These are the even-Z elements thorium (90) and uranium (92). Similar properties we observe for nuclei, which have odd or even neutron numbers. We call those nuclei which all have the same number of neutrons 'isotones' in analogy to 'isotopes' which all have the same number of protons. You may prove the regularity with the help of Figure 1.5. But you need to look carefully, like scientists would.

4

Elements made with light ion beams from 'small' cyclotrons

Ernest Orlando Lawrence conceived the idea of the cyclotron in 1929 when he went to the University of California at Berkeley. There, together with his student M. Stanley Livingston, he built a cyclotron for accelerating hydrogen ions (protons) up to an energy of 13 keV. Although this model was only 4.5 inches in diameter, it demonstrated the principal of acceleration. A second cyclotron followed, able to accelerate protons up to 1.2 MeV, high enough to cause nuclear disintegration. In 1936, Lawrence was able to fund the establishment of what would become famous the world over, the Radiation Laboratory with its 60-inch cyclotron in which several new elements were synthesised; it was the sixth machine in the series. Lawrence received the Nobel Prize for Physics in 1939 for his invention of the cyclotron.

It was one of Lawrence's cyclotrons which produced the first element not found in nature, technetium ($Z = 43$). The chemistry of this hitherto unknown element had been predicted on the basis of the Periodic Table. It was discovered in 1937 by Carlo Perrier and Emilio Segrè in Italy; they analysed a sample of molybdenum ($Z = 42$) which was irradiated in Berkeley with deuterons and sent to Italy for further study.

Curium (element 96)

The first transuranic element produced at the Berkeley 60-inch cyclotron was plutonium, element 94; like technetium it was synthesised with a deuteron beam. The third was curium, element 96, found when small amounts of ^{239}Pu were irradiated with 32 MeV helium ions:

$$^{4}He + ^{239}Pu \rightarrow ^{242}Cm + n$$

'The irradiation was done in summer 1944 in the Berkeley 60-inch cyclotron after which the material was shipped to the Metallurgical Laboratory at Chicago for chemical separation and identification' (Seaborg and Loveland 1990). There, Glenn T. Seaborg, Ralph A. James and Albert Ghiorso were able unambiguously to identify the new α-emitting nuclide ^{242}Cm. The crucial step in their identification was the measurement of the known

Box 4.1 The cyclotron

A cyclotron consists of two hollow semicircular electrodes called
'dees' after their shape like the capital letter D. These are mounted
back to back in an evacuated chamber, separated by a narrow gap.
The chamber is placed between the poles of an electromagnet and an
electric field, alternating in polarity, created in the gap by a radio-
frequency oscillator.

The particles to be accelerated are formed as ions (charged
atoms) in the gap near the centre of the device, where the electric field
propels them into one of the dees (Figure 4.1). There the magnetic
field guides them in a semicircular path. By the time they return to the
gap, the electric field has reversed, so they are accelerated into the
other dee. Although the speed of the particles and the radius of their
orbit increase each time they cross the gap, as long as the mass of the
particles and the strength of the magnetic field remain constant, these

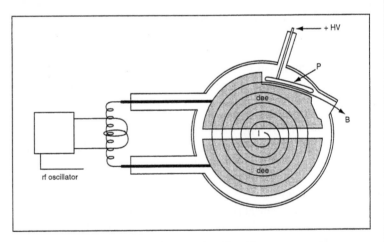

Figure 4.1 Essential parts of a cyclotron, not including the magnet, showing
dees, resonant circuit with an rf power source and deflector plate P
connected to a positive high voltage source (+HV). The path of the
ions from the ion source at the centre I to the point of emergence at
B is shown schematically. The accelerating electric field acts only in
the gap between the dees, not inside the dees itself. The pancake-
like arrangement is put into a flat cylindrical vacuum chamber
which is then moved between the poles of a strong magnet. The
direction of the magnetic field is perpendicular to the drawing. The
figure was taken from Segrè (1965).

crossings occur at a fixed frequency to which the oscillator can be adjusted.

The result is a stream of particles of high speed which is, however, not continuous but proceeds in bunches with periods determined by the travelling time of the particles through the dees. After acceleration, the bunches of particles can be deflected out of the dees to targets in various detector set-ups around the instrument.

α-emitter ^{238}Pu as daughter product after the α-decay of ^{242}Cm. After careful chemical study they found ways of separating curium from its parent plutonium and confirmed their conclusions.

They decided to call element 96 'curium' (Cm) after Pierre and Marie Curie, pioneers in the study of radioactivity. In the Periodic Table it lies above gadolinium; the chemical similarity of the two elements derives from the fact that the outer ten electrons occupy orbits which have almost identical properties.

Berkelium and californium (elements 97 and 98)

Elements 97 and 98 were produced in the Berkeley 60-inch cyclotron by irradiation of ^{241}Am and ^{242}Cm with helium isotopes:

$$^4He + \,^{241}Am \rightarrow \,^{243}Bk + 2n$$
$$^4He + \,^{242}Cm \rightarrow \,^{245}Cf + n$$

The final irradiations took place in December 1949 and February 1950, respectively. The work on element 97 was done by Stanley G. Thompson, Ghiorso, and Seaborg, joined by Kenneth Street Jr for element 98.

The most important prerequisite for the synthesis of these two elements was the manufacture of sufficiently large amounts of americium and curium to serve as target materials. Americium was prepared in milligram amounts by the intense neutron irradiation of plutonium over a long period of time, while curium was made in microgram amounts by equally intense neutron bombardment of some of this americium.

(Seaborg and Loveland 1990)

The second difficulty was to separate the newly produced elements from the target material. The kinetic energy of fusion products produced in reactions with helium is so small that most of the products are stopped inside the target.

The separation required sophisticated chemical separation steps which had to be worked out before the experiment; it took the group three years.

> Element 97 was called berkelium (Bk) after the city of Berkeley, California where it was discovered, just as the name of its rare earth analogue, *terbium*, was given a name derived from Ytterby, Sweden where so many of the early rare-earth minerals were found. Element 98 was named californium (Cf) after the university and state where the work was done.
>
> (ibid.)

Mendelevium (element 101)

'The discovery of *mendelevium* (element 101) was one of the most dramatic in the sequence of transuranic element synthesis in Berkeley. It marked the first time in which a new element was produced and identified one atom at a time' (ibid.).

Using neutron irradiation of plutonium in the Materials Testing Reactor in Idaho, the Berkeley group had by 1955 prepared an equilibrium amount of about 10^9 atoms of ^{253}Es, an α-emitter with a half-life of only 20 days. Helium ions, from the 60-inch cyclotron were again used for the irradiation with flux increased to some 10^{14} ions/second. The reaction expected to produce element 101 was:

$$^4\text{He} + {}^{253}\text{Es} \rightarrow {}^{256}101 + \text{n}$$

Al Ghiorso, in a 'back of the envelope' calculation (ibid.) during an aeroplane flight, estimated the number of atoms to be expected. It turned out to be just one atom of element 101 for three hours' irradiation! As you can imagine,

> there was desperate need both for new techniques and for some luck. The first new technique involved separation of the element 101 by the recoil method [Figure 4.2]. An invisibly thin layer of einsteinium was placed on a gold foil. The helium-ion beam was sent through the back of the foil so that the atoms of element 101, recoiling through a vacuum due to the momentum of the impinging helium ions, could be caught on a second thin gold catcher foil. The catcher containing recoil atoms, relatively free of the einsteinium target material, was dissolved and used for later chemical operations.
>
> (ibid.)

'The earliest experiments were confined to a search for short-lived, α-emitting isotopes. No such activity was observed attributable to element

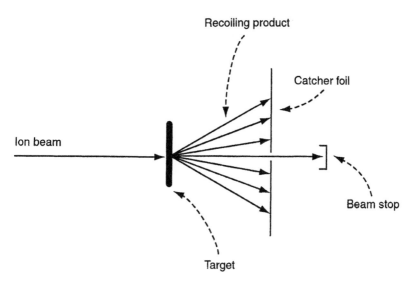

Figure 4.2 Recoil catching technique. The beam current is measured using a current meter connected to the beam stop. The figure was taken from Seaborg and Loveland (1990).

101' (ibid.). Nevertheless, the experiments were continued and, in one of the subsequent overnight irradiations, two large pulses were observed in the electronic detection apparatus resulting from spontaneous fission.

'The definitive experiments were performed in an all-night session on February 18th, 1955' (ibid.). Using other known reaction products, and the decay of the target material itself, it was possible to define the positions in which the elements were eluted from off an ion-exchange column.

> A total of 5 spontaneous fission counts were observed in the element 101 position, while 8 were also detected in the element 100 position. The four pulses occurred at 01:15, 01:37, 02:40 and 10:35 on February 19th. Al Ghiorso and Bernie Harvey made a note directly on the recording chart: 'This experiment conclusively proves the chemical identification of element 101.'
>
> (ibid.)

The decay sequence they discovered was:

^{256}Md + e$^-$ → ^{256}Fm + ν (this is an electron-capture decay),

^{256}Fm → ff$_1$ + ff$_2$ (spontaneous fission; 'ff' = 'fission fragment').

The discovery team (Albert Ghiorso, Bernard G. Harvey, Gregory R. Choppin, Stanley G. Thompson and Glenn T. Seaborg) suggested the

> name *mendelevium* for element 101 in recognition of the role of the great Russian chemist Dmitri I. Mendeleev who was the first to use the periodic system to predict the chemical properties of undiscovered elements. The proposed chemical symbol (Mv) was later changed to Md by the International Union of Pure and Applied Chemistry.
>
> (ibid.)

5

Hot fusion and controversial discoveries: elements 102 to 106

With mendelevium, element synthesis using helium-ion beams and reactor-generated actinide targets came to an end. In order to proceed to element 102, one would need fermium targets which cannot be made. To go beyond element 101, projectiles able to add more than two protons in a fusion reaction were required, necessitating ions heavier than helium. Such beams became available in 1957 at the HILAC (Heavy Ion Linear Accelerator) in Berkeley and in 1960 at the U-300 heavy-ion cyclotron of the Joint Institute for Nuclear Research (JINR) in Dubna, north of Moscow.

Each laboratory pursued element synthesis with its own specific strategy. The Berkeley group of Ghiorso and co-workers, having access to heavy actinide targets, used lighter projectiles than the Dubna workers led by Flerov who relied on light actinide targets. For example, to make element 104 the Berkeley group applied the reaction

$$^{12}C + {}^{249}Cf \rightarrow {}^{257}104 + 4n$$

while the Dubna group used

$$^{22}Ne + {}^{242}Pu \rightarrow {}^{259}104 + 5n$$

Projectile energies of 5–6 MeV/nucleon are needed in order to overcome the electric repulsion acting between the positively charged projectile and target nucleus. The experts introduced the term 'Coulomb barrier' for this effect. It is like crossing a mountain pass: you have to have enough energy to get to the top; coming down the other side is easy. Thus, the kinetic energy of the projectile must be high enough to overcome the mutual repulsion between it and the nucleus – once that has been done, attractive nuclear forces start to act and suck in the projectile, so forming a compound nucleus. The latter is not identical with the nucleus which finally survives the reaction because binding energy is set free which heats up the compound nucleus. Cooling then takes by evaporation of neutrons and emission of gamma rays, just as hot water cools by evaporation of vapour and radiation of heat.

The temperature or excitation energy in nuclear physics is measured and expressed as MeV. The evaporation residue eventually reaches an energy minimum which is *the ground state of the nucleus*. The whole fusion process, including the time for cooling down, is over in an unbelievably short time, one millionth of a billionth (10^{-15}) of a second.

Evaporative cooling is very effective for nuclei, however, for heavy nuclei a competitive process exists. Heavy nuclei carry so many protons, and are therefore highly charged, that they are very fragile and tend to break (fission) in two. If the heavy nucleus is hot, fission is enhanced. Heavy compound nuclei have only a limited period of cooling down and surviving by neutron evaporation, because the fission of hot compound nuclei is much faster and, therefore, the main limitation in the production of heavy elements. Nevertheless, five elements have emerged from hot fusion reactions: nobelium (No; $Z = 102$), lawrencium (Lr; 103), rutherfordium (Rf; 104), dubnium (Db; 105) and seaborgium (Sg; 106).

Several discoveries became points of controversy between Berkeley and Dubna. Different names for the same element were even used in East and West. This led the International Unions of Pure and Applied Chemistry (IUPAC) and of Physics (IUPAP) to set up in 1985 a 'Transfermium Working Group' (TWG) to establish criteria for the discovery of new elements and to prove existing claims. It turned out that, to handle these issues, discovery profiles incorporating multiple contributions made at various times were often appropriate. It was concluded that the discoveries of elements 103 to 105 should be shared between the Berkeley and Dubna groups. After further discussions in the scientific community, the IUPAC eventually adopted the names and symbols used here.

At the end of the actinide series, half-lives drop down to the range of seconds, too short for 'off-line' chemical identification methods. Nonetheless, it was possible to overcome this difficulty with 'on-line' chemistry (see Box 5.1) as shown for element 104 (the first transactinide element) in pioneering experiments by Ivo Zvara and his group in Dubna; their contribution was heavily weighted in their favour in the TWG's discovery profile for this element.

The Berkeley group replaced chemical identification with purely physical techniques to establish generic relations between a new element and well known decay products. Figure 5.1 shows the set-up, the vertical wheel in its most advanced version used in 1974 for the discovery of element 106, seaborgium. The group of co-discoverers is shown in Figure 5.2. Seaborgium was produced by the reaction:

$$^{18}O + {}^{249}Cf \rightarrow {}^{263}Sg + 4n$$

The isotope ^{263}Sg with a half-life of 0.9 sec was identified by observation of the previously known ^{259}Rf and ^{255}No as daughter and granddaughter products.

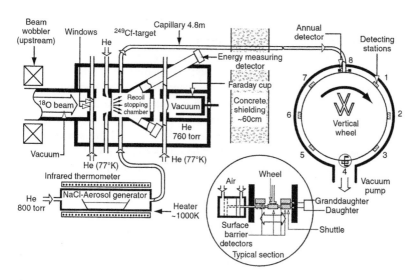

Figure 5.1 The VW (vertical wheel) set-up used in 1974 by a Berkeley–Livermore collaboration, Albert Ghiorso *et al.*, E. Kenneth Hulet *et al.*, for the identification of element 106, seaborgium. The 0.9-second ^{263}Sg from the ^{18}O + ^{249}Cf → ^{263}Sg + 4n reaction was shown to decay by α-particle emission to the known three-second ^{259}Rf, which in turn decayed to the known three-minutes ^{255}No granddaughter. Reaction products recoiling from the target were stopped in helium seeded with NaCl aerosols. The helium jet carried the reaction products attached to the aerosol particles through a long capillary to the set-up where they were deposited on a wheel. An annular silicon detector analysed the α-particle spectra during collection. The wheel was rotated stepwise to seven detector stations (insert), each comprising two movable silicon surface-barrier detectors below and two stationary detectors at both sides of the wheel. One of the movable detectors faced the rim of the wheel to examine the α-particle spectrum of the deposit; it also collected daughter products transferred from the rim by α-recoil. Such daughter and granddaughter products were detected after moving this detector opposite to a stationary detector, while the second movable detector took over the position at the rim. The figure was taken from Ghiorso *et al.* (1974).

Seaborgium is currently the last heaviest element but two whose chemical properties are known, although it took more than two decades for a American–German–Russian–Swiss collaboration to reveal them. It became a paradigm of how the Periodic Table can be used to order and correlate chemical properties. The difficulties resulted from experimental problems: chemical characterisation needed methods that could work with single

Box 5.1 'Off-line' and 'on-line' chemistry

Chemical separation needs time, enough material, a well equipped laboratory and, of course, an experienced chemist. With all of these, individual reaction products can be isolated and their atomic numbers established. The process of separation is quite separate, both in time and place, from the actual creation of the new element: the chemistry is said to be *off-line*. An extreme example of this is the search for superheavy elements in nature. Short half-lives can be investigated off-line using the so-called SRAFAP method, 'Students Running As Fast As Possible'.

If the half-lives drop below a few minutes, the chemical 'laboratory' has to move *on-line*, close to the place where the reaction takes place. If this is a nuclear reactor or an accelerator, the chemistry has to be automated because living chemists cannot work in such a hazardous environment; the shortest half-life limits achievable are some tenths of seconds.

atoms, were fast enough to handle short-lived nuclides and could be run over extended periods to accumulate statistically significant numbers of events. The work was helped greatly by the earlier discovery by a Dubna–Livermore (the latter a major California laboratory) collaboration of two relatively long-lived α-decaying isotopes, ^{265}Sg and ^{266}Sg. So far, the chemical properties of seaborgium are in agreement with its placement in group VI of the Periodic Table.

Figure 5.2 The co-discoverers of element 106, seaborgium (Sg) at the Heavy Ion Linear Accelerator building of the Lawrence Berkeley Laboratory at the time of discovery in 1974. From left to right: Matti Nurmia, Jose R. Alonso, Albert Ghiorso, E. Kenneth Hulet, Carol T. Alonso, Ronald W. Lougheed, Glenn T. Seaborg and J. Michael Nitschke. The photograph was taken from Seaborg (1995).

6

Cold fusion and 'big' cyclotrons

Chemical reactions can be exothermic or endothermic: energy as heat is either liberated or required. How much energy – and in which direction – depends on the chemical bonds involved. Furthermore, a reacting chemical system can, of course, easily be heated or cooled from without.

A reacting nuclear system behaves in a similar fashion although, unfortunately, we cannot cool it from the outside. Cooling would be highly desirable because at the minimum beam energy necessary for a reaction, the nuclear reaction is still exothermic: heat is emitted and the reaction product (the compound nucleus) gets hot. Difficult though cooling may be, further heating is easy: just increase the energy of the projectile ion which creates heat in the form of excitation energy of the nucleons within the nuclei.

A remarkable variation of this excitation energy at minimum beam energy occurs if we make a particular compound nucleus using different projectile target combinations. The distribution of excitation energies in the case of ^{262}Sg as compound nucleus is shown in Figure 6.1. A maximum value of 45 MeV is reached for the combination ^{20}Ne + ^{242}Cm and a minimum of 18 MeV for ^{54}Cr + ^{208}Pb. The reason for this variation lies in the different binding energies of the reacting nuclei: the binding energy for ^{54}Cr + ^{208}Pb is less than that for ^{20}Ne + ^{242}Cm (remember that the binding energies of stable nuclei are negative. Therefore less binding energy means stronger binding).

In the first case, the excitation energy of the compound nucleus after fusion is relatively high and four neutrons plus gamma rays have to be emitted to cool the nucleus down to the ground-state. In the second case, the reaction is cold and the emission of only one neutron plus gamma rays is sufficient to reach the ground-state.

However, the number of emitted neutrons alone is not a sufficient guide for deciding which type of reaction gives higher yields. This was a matter of great uncertainty and the problem could not be solved theoretically. So far, only hot fusion had been examined experimentally because only light projectiles were available from accelerators. It was Yuri Tsolakovich

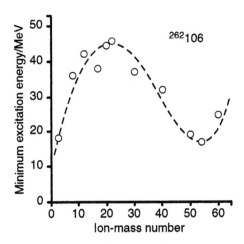

Figure 6.1 Minimum excitation energies of various fusion reactions all resulting in the compound nucleus $^{262}106$. Excitation energies not less than about 45 MeV are obtained in reactions using actinide targets and projectiles with mass numbers $A \approx 20$ (hot fusion). Minimum excitation energy of about 18 MeV is obtained with ^{208}Pb target and ^{54}Cr projectiles (cold fusion). The figure was taken from Oganessian (1974).

Oganessian at Dubna who first considered cold fusion as an alternative path for the synthesis of the heavy elements.

The breakthrough – cold fusion worked well

A number of experiments were carried out in Dubna using the three-metre diameter cyclotron U-300 operating since 1960. Beams from ^{40}Ar to ^{76}Ge were used, and targets of lead and bismuth, but until 1973 the experiments were not successful. The negative results were explained by the high fissility, i.e. the tendency of the reacting system to re-separate on the way to fusion. Quasifission is the expert terminology. The fissility increases with the product of the proton numbers of the projectile and target. In the case of Ne + U the product is $10 \times 92 = 920$; with Ca + Pb it is $20 \times 82 = 1,640$; in both cases, element 102 is the fusion product. We see a tremendous increase of the fissility for the reactions with lead which led people to believe that cold fusion would never work. This assumption was supported, of course, by the negative results of the early experiments but, as it soon turned out, these were not sufficiently sensitive.

In 1973 the detection methods were improved and the reaction ^{40}Ar + ^{208}Pb for synthesising fermium (element 100) was investigated. 'The results of the experiments surpassed all expectations', wrote Georgi Flerov in a review article (Flerov and Ter-Akopian 1987: 215):

> instead of the usual several tens of atoms, their yield amounted to several hundreds and even thousands of atoms in a one-day experiment. The largest cross-section characterised the reactions involving the evaporation of 2 and 3 neutrons from the compound nucleus rather than the reactions accompanied by the evaporation of 4–5 neutrons. The latter are typical for hot fusion. This fact is convincing evidence that the use of 'magic' lead nuclei (those with closed shells for the protons and neutrons) as target material should lead to the formation of slightly excited (cold) compound nuclei.

Nevertheless, for many years the results of the Dubna cold fusion reactions were not accepted by the Berkeley scientists.

A typical experimental arrangement is shown in Figure 6.2. The ion beam was incident tangentially to the surface of a thin walled aluminium cylinder covered by the target material. The diameter of the cylinder was 10 cm. The whole device was installed inside the cyclotron, near the outer circumference but still between the poles of the magnet. Only there was the beam intensity high enough to make a search for new elements successful. The cylinder was mounted so that the axis of rotation was vertical along the magnetic lines inside the cyclotron. The cylinder could be rotated at various speeds (with a maximum of 5,600 revolutions/minute) to measure lifetimes over a wide range, with the shortest one only a few milliseconds.

The target layer (Tl, Pb or Bi) had a thickness of only 1.8–2.7 micrometres although the effective layer for the beam was several times thicker due to the tangential irradiation. Along its pass through the target material, the beam loses energy so that a whole range of beam energies was covered during one irradiation. This is an important practical aspect because the maximum yield for the production of an element happens only at a well defined beam energy which is, however, usually not accurately known. The fusion products were found close to the target's surface; they were stopped almost immediately after the reaction had taken place. If the heavy nucleus produced decayed by fission, one of the two fission fragments could always easily escape from the target layer.

The escaping fission fragments were detected using thin foils of a mineral (muscovite mica). The regular structure of the crystals of this mineral is destroyed when a heavy particle like a fission fragment is stopped in it. After etching, the traces of the particles become visible under a microscope. Their length is only about 15 micrometres. The foils were mounted in a fixed

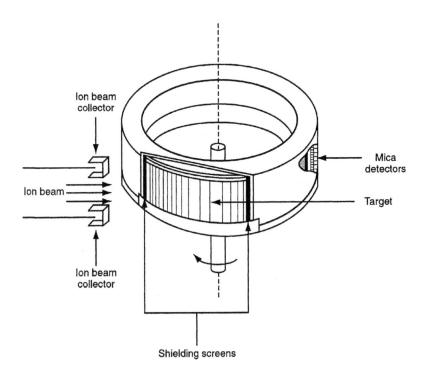

Figure 6.2 Schematic view of the rotating target cylinder used in Dubna for irradiation inside the cyclotron. The mica fission-track detectors were fixed around the circumference. The figure was taken from Oganessian *et al.* (1976).

position around the rotating cylinder enabling some 90% of all fission events to be detected. An example of how the tracks are distributed on the foils is shown in Figure 6.3. The half-life was determined from the decreasing density of the tracks with distance from the irradiation area: when the rotation speed of the cylinder is known, the location can be easily converted into a time scale.

The device looks simpler than it really was because it had to rotate at high speed in vacuum and in a strong magnetic field. High currents ('eddy-currents') are induced when metals rotate in magnetic fields. As these currents tend to slow down the rotation, a special design was needed to reduce them to a minimum. Intensive cooling of the cylinder protected targets like Tl, Pb, and Bi which have low melting points. Cooling was achieved by two heat-removal circuits with a liquid metal (an indium–gallium alloy) and water as coolants in the first and second circuit, respectively. Only then could the arrangement operate at very high beam intensities.

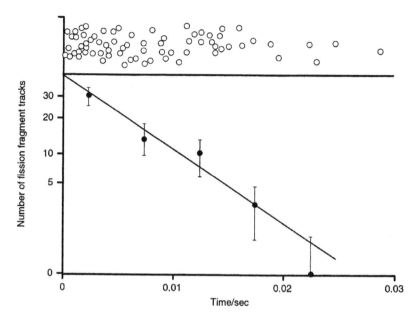

Figure 6.3 Time distribution of spontaneous fission tracks in the reaction $^{50}Ti + ^{208}Pb \rightarrow ^{258}Rf^*$ (the star symbol is used to indicate an excited compound nucleus). The tracks became visible under a microscope after etching the mica fission-track detectors which were fixed around the rotating cylinder (see Figure 6.2). In the lower part of the figure the decay curve is plotted after subtraction of the contribution from a long lived activity with a half-life of several seconds. For the nucleus ^{256}Rf (the resulting nucleus after evaporation of two neutrons) a half-life of five milliseconds was deduced from the decay curve. The most recent value is 6.2 milliseconds. The figure was taken from Oganessian *et al.* (1975).

New results – unexpected fission half-lives

Another important milestone in heavy element research was achieved at Dubna in 1974–5. Using the rotating cylinder and fission fragment detectors, Oganessian and his co-workers investigated isotopes of element 104 by cold fusion. They used a beam of ^{50}Ti ions and lead isotopes of mass numbers 206, 207 and 208 as targets. What they found surprised them: the new nuclide they got (^{256}Rf) had a half-life of about five milliseconds, 10,000 times less than expected from the extrapolation of the data established in Berkeley by Ghiorso.

Figure 6.4 shows the trend of the half-lives for the elements from curium to 104; a sharp half-life maximum at neutron number 152 is clearly visible. Extrapolating the shape of the curves beyond element 102, one

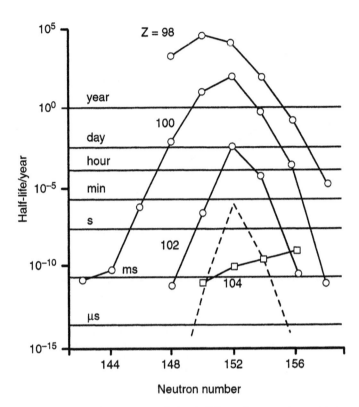

Figure 6.4 Systematics of spontaneous fission half-lives of even–even nuclei at neutron number N = 152. Experimental data are shown by open circles and squares connected by full lines. The dashed line shows the extrapolation by Ghiorso (1969) for the element 104 isotopes. The figure was taken from Flerov and Ter Akopian (1987).

would expect a half-life of minutes for element 104 but instead it was no more than milliseconds. On the other hand, the isotope with neutron number 156, ^{260}Rf, had a half-life a million times longer than expected. The Berkeley group was very sceptical.

The result was very important for the future of heavy element research, because it indicated a drastic change in the systematic of fission half-lives. If confirmed, it could mean that all elements beyond element 102 could fission and, even more significant, that the fission half-lives could become shorter and shorter as predicted by theoretical calculations. This was bad news: while fission is easy to detect because of the strong signals from the detectors, it is difficult to determine the element and mass number and so to identify a nucleus which decays just by fission. Second, if the half-lives dropped below a millionth of a second, there would be no chance at all of identifying those short-lived elements. The new results from Dubna therefore created a lot of excitement.

Oganessian and his group had a simple qualitative explanation for the short half-life of ^{256}Rf, overlooked for a while but soon confirmed by theoretical calculations. For a better understanding of what follows, first read Box 6.1 'Tunnelling – how particles pass through walls and mountains'.

Box 6.1 Tunnelling – how particles travel through walls and mountains

Movies sometimes seem to show an actor going through walls. For a while, this leads to surprising and funny situations but the actor is usually happy when he comes back to a normal existence. If normal people try to go through walls, the wall will reflect them back into reality, usually painfully! To avoid this we must go round the wall or, if that is not possible, jump over it. In jumping over it, we convert kinetic energy in the act of jumping into potential energy at the top of the wall, converted in turn back into kinetic energy coming down the other side of the wall.

This is more complicated in the microcosm of atoms, nuclei and elementary particles. There, walls made of the fields of electric and nuclear forces exist as potential barriers. (You can get an idea of how potential barriers work when you are playing with strong magnets, putting one below and one above a tabletop.)

Let us consider the emission of alpha particles from a nucleus. To my knowledge the first case of tunnelling through a potential barrier was correctly described by George Gamow.

Figure 6.5 shows the kinetic and potential energy of an alpha particle inside and near a nucleus as a function of the distance from the centre of the nucleus (at $r = 0$). The barrier is formed by the attractive nuclear force resulting in negative energies in our energy–distance diagram and the repulsive electric force – repulsive because the nucleus and alpha particle are both positively charged. Inside the nucleus (at distances r less than R_{in}) the strong attractive nuclear force dominates. All energy levels of negative energy, the bound states, are filled with nucleons. The alpha particle emerges from a state of positive energy. This is the case for alpha unbound, radioactive nuclei. If the alpha particle were neutral, the barrier would not exist and the particle would be immediately emitted. However, as the alpha particle is

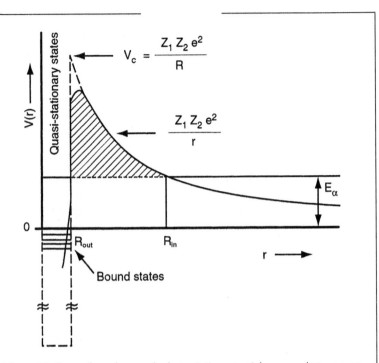

Figure 6.5 Sum of nuclear and electrostatic potential energy between an
α-particle and the daughter nucleus as a function of the distance
from the centre of the (daughter) nucleus at r = 0. Although the α-
particle is energetically unbound, it is nevertheless prevented from
escaping from the nucleus by the potential barrier (hatched area)
which is formed from the sum of potentials based on nuclear and
electrostatic forces. We say the α-particle is in a 'quasi-stationary'
state. A certain quantum mechanical probability exists for the α-particle to traverse the potential barrier and escape from the nucleus
with increasing speed due to the Coulomb repulsion. The maximum
speed is determined by the energy E_α. The dashed curve shows the
potential assuming that the nucleus has a sharp surface at the radius
R (simple model, square well potential), whereas the drawn curve
shows the more realistic case assuming a diffuse nuclear surface.

positively charged, it is repelled by the surrounding electric field of the
positively charged nucleus which traps it inside the nucleus. Because
the shape of the barrier outside the nucleus is formed by electric
forces, it was given the name *Coulomb barrier* after the physicist
Charles Augustin de Coulomb who, in 1785, discovered the law

governing electric forces acting between two charged spheres. Surprisingly, this law remains valid without changes in the regime of such tiny systems as elementary particles.

In our classical world, the alpha particle could escape from the nucleus only if kinetic energy inside the nucleus were higher than the potential energy maximum. In Figure 6.5, the kinetic energy is energy of the alpha particle above zero. In microcosm, however, the alpha particle behaves like a wave and, as a wave, it can penetrate into the forbidden region below the potential barrier. This effect is similar to the penetration of long radio waves below the surface of the ocean which allows contact to be maintained with submarines. In this way, the alpha particle wave can penetrate so deeply into the barrier that it appears again as the alpha particle on the other side. If this happens, the alpha particle is rejected by the electric force and it recovers back its kinetic energy with increasing distance from the nucleus. It is this kinetic energy which is very specific for the alpha decay of nuclei and which we measure with our detectors. The process was called *tunnelling* and the probability of tunnelling is closely related to the half-life.

It was Gamow who first understood the law for the probability of an alpha particle traversing the barrier by tunnelling. This probability decreases with an increase of the area formed by the cross-section through the mountain (i.e. the barrier) above the tunnel. The area is drawn hatched in Figure 6.5. The higher and wider the barrier, the smaller is the probability for tunnelling and the longer is the half-life of the nucleus for alpha decay.

There is also a barrier in the case of spontaneous fission but this is more complicated due to the high mass and charge of the fission products and one must also include the variation of the nuclear potential when the nucleus deforms. These deformations are shown in Figure 3.3. The fission barrier is accordingly usually plotted as function of the deformation and not as function of the distance, as in the case of alpha decay.

Finally, both alpha decay and fission can be inverted to describe fusion with light and heavy projectiles. In the case of fusion, the Coulomb barrier must be crossed but now it is from the outside. This means that the kinetic energy of the projectile must be high enough, at least as high as the barrier, *the fusion barrier*. However, due to tunnelling, fusion is also possible at lower energy, the so-called *sub-barrier* energy. But the probability for fusion decreases because the lower the energy, the higher the fusion barrier. The positive aspect is that also the excitation energy of the compound nucleus becomes lower which, again, lowers the probability for its fission during the cooling down phase.

From Box 6.1 we can see that, for nuclei and elementary particles, there is a probability of passing through mountains instead of consuming energy by first clambering high (for which we use an accelerator) and then flying over the top; the process is called *tunnelling*. The probability of tunnelling depends not only on the length of the tunnel but also on the height of the 'mountain' (energy barrier) above the tunnel and how this barrier is structured in detail. In general, the higher and wider the barrier above the tunnel, the lower is the probability of a particle crossing it.

To return to Oganessian's explanation, Figure 6.6 shows a cross section through the mountain for the fission of ^{252}Fm (Z = 100) and some neighbouring nuclei. ^{252}Fm is the isotope with the longest fission half-life located at neutron number N = 152 (see Figure 6.4). The ground-state of ^{252}Fm is the valley at the left of the hatched peak area. This energy determines at which altitude the fission tunnel through the mountain is located (dashed line). In the case of ^{252}Fm the mountain above the tunnel is high and also wide because of the second smaller peak at the right side. This makes the probability for tunnelling small or, correspondingly, the fission half-life long. In professional language the mountain is called the *fission barrier* and ^{252}Fm has a *double humped* fission barrier. A double humped fission barrier still exists for ^{254}No, although it is less pronounced. Now consider ^{256}Rf. There, the second hump almost disappeared. The result in this case is that the tunnel lies above the second hump and the fission barrier for ^{256}Rf is much narrower. That is why the fission half-life of ^{256}Rf is 10,000 times less than expected from the data measured in Berkeley.

The Dubna result was later confirmed, the most recent value for the half-life of ^{256}Rf being (6.2 ± 0.2) milliseconds. Theoreticians also substantiated Oganessian's qualitative picture: it could be shown that nuclei with neutron number N = 152 are more stable than the neighbouring isotopes and that there does indeed exist a second hump in the fission barrier which significantly influences the fission half-life.

Oganessian's explanation appeared to convince the people in Berkeley. After correction of the data for fission half-lives, many (perhaps most) experts believed that fission would be the dominant decay mode for the

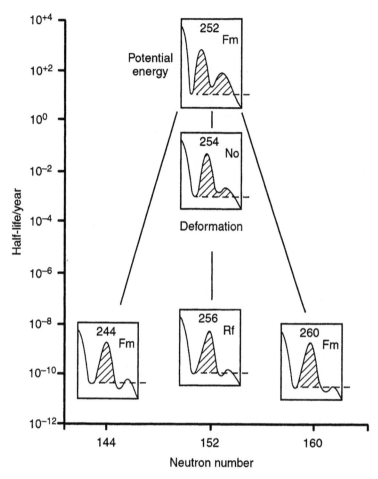

Figure 6.6 Illustrative representation of the various fission barriers for even–even nuclei with neutron number near N = 152. The wide and double-humped fission barrier in the case of ^{252}Fm and ^{254}No is responsible for the long partial fission half-life of those nuclei. The figure was taken from Oganessian (1975), modified.

elements beyond 104, a conclusion supported by theoretical calculations. It was predicted that half-lives should decrease further with increasing element number. So fission-track detectors became the method of choice at Dubna in the search for elements 106 and beyond. However, this turned out to be an unfortunate choice.

The U-400 cyclotron at Dubna

After cold fusion using targets of lead and bismuth was established as a successful method for the synthesis of heavy elements, the Dubna people wanted to go a step further and considered the synthesis of elements beyond 105. That was easier said than done because the main prerequisite, a more powerful accelerator, was not available. Certainly, it was asking a lot of the funders to build a huge new machine on the basis of vague assumptions. Nevertheless, Oganessian was able to convince them and a powerful new heavy ion cyclotron accelerator was built in Dubna during the 1970s.

When designing a new accelerator for heavy ions, the first step is to think about the maximum velocity which the accelerated ions have to reach and also about the heaviest ion which one wants to accelerate. The minimum velocity which projectiles need to impact the target nucleus to yield an element in the region $Z = 100$–110 is about 10% the velocity of light, or 30,000 km/sec, a value about the same for all combinations of projectile and target: for very asymmetric systems using light projectiles as well as for the heaviest possible targets, like $^{22}Ne + ^{248}Cm$ for the synthesis of element 106, and for almost symmetric combinations like $^{136}Xe + ^{130}Te$. This reaction would also result in element 106 but it had not yet been explored experimentally. Some physical, together with technical, considerations argued against symmetric systems for the synthesis of heavy elements as we will see later. However, physics is always good for surprises.

The aim of the Dubna group was the construction of an accelerator able to accelerate ions as heavy as ^{136}Xe up to 10% of the velocity of light. The velocity which can be reached in a cyclotron for projectiles of specified mass and charge is related in a very simple way to the parameters of the cyclotron. The maximum velocity increases with the magnetic field strength and, of course, with the cyclotron radius; it also increases with the charge of the ion and, because of the centrifugal forces, decreases with increasing mass. However, as there is a limit to the magnetic field obtainable with the iron and the conducting coils of magnets, the use of heavy ions for heavy element synthesis requires a cyclotron with a larger radius than that of the one they had.

The choice was made to build a cyclotron of 400 cm diameter. From this comes the name, U-400. The 'U' stands for the Russian word '*Uskoritel*' which means accelerator. The total weight is 2,100 tons (mainly iron) and an electric current of 2,500 Amperes is needed to create the magnetic field. The new cyclotron started to operate in 1981. With it, ions as heavy as ^{64}Ni could be accelerated up to energies high enough for fusion reactions. For a beam of ^{136}Xe, however, neither the energy nor the intensity was sufficient for the synthesis of heavy elements.

At that time, the experimental programme at Dubna using the new cyclotron was tremendous. Many new data on the fission properties of known heavy elements were measured. Attempts were also made to produce the new elements 107, 108, 109 and even 110 and 111. However, the drawback of the Dubna experiments was, as we have already mentioned, that they were sensitive only for fission. Thus, the data were not unambiguous and it was not easy to assign the measured fission activities to a certain nucleus.

What was still missing in Dubna was an instrument for the fast and efficient separation of the reaction products after fusion and a high sensitivity of the experimental set-up not only for detecting fission but also for alpha decay. Such a separator and detector system had already been constructed in the early 1970s at the new research centre for heavy ions in Darmstadt near Frankfurt in Germany. I will come back to the developments there in the next chapter. Before doing so, however, I want to describe one method which allows for an indirect proof that heavy elements may decay by emission of alpha particles: the chemical separation of daughter decay products which have long half-lives and, therefore, highly sensitive off-line chemistry can be used. As well as searching for fission with the fission track detectors, chemistry was extensively used in Dubna until the end of the 1980s.

Radio-chemical separation

Some α-decay chains of heavy elements end in well-known isotopes and some of these isotopes have half-lives from a few hours to several days, long enough to perform a clean off-line chemical separation of the daughter elements from the bulk of other reaction products. After chemical separation the probe is put in front of alpha or fission detectors. These measure the activity while the measured decay energy allows determination of the isotopic composition of the sample. Applying this procedure for heavy elements, information about the parent nucleus could be obtained even if that nucleus could not be measured directly, perhaps because the experimental conditions did not enable the detection of alpha decay or short half-lives. The method can, of course, be applied only if the parent nucleus decays by alpha emission but, on the other hand, a positive result is also an indication of alpha decay of the parent nucleus. This was the basic idea used by the Dubna scientists to obtain information on the decay properties of elements 107 and 109. Both the intensive internal beam of the cyclotron and the reliable target technology of the cooled rotating cylinder were available. The hope was that the data would give information about the production cross-section for the heaviest, and at the time still unknown, elements and whether they decayed by alpha emission or fission.

For the production of the elements 107 and 109, one more advantage was the variety of natural projectile isotopes. Starting with the most neutron-rich isotope of titanium, ^{50}Ti, there exists a series of projectiles which differ by additional alpha particles: ^{54}Cr and ^{58}Fe. In the irradiation of ^{209}Bi targets with these projectiles, the following compound nuclei are produced: ^{259}Db (dubnium, Z = 105), ^{263}Bh (bohrium, 107) and ^{267}Mt (meitnerium, 109). After emission of one neutron – we are still dealing with cold fusion reaction – the end products (the evaporation residues) are: ^{258}Db, ^{262}Bh and ^{266}Mt. Each differs from the next by one alpha particle. Assuming that these isotopes are α-emitters, the 35 hour half-life isotope ^{246}Cf (98) would eventually be produced after α decay of the nucleus ^{254}Lr (103) (which was already known), electron capture decay of ^{250}Md and another alpha decay of ^{250}Fm (100). The decay chain is shown in Figure 6.7.

In a series of experiments the Dubna scientists started with the production of element 105, then 107 and 109 using projectile beams of ^{50}Ti, ^{54}Cr and ^{58}Fe, respectively. The daughter isotope ^{246}Cf was readily separated radiochemically after an irradiation period of one to two days and identified by its α-decay spectrum. The line intensities decreased by a factor of 10 from ^{258}105 to ^{262}107 and by a factor of almost 100 from ^{262}107 to ^{266}109. These results were the first serious indications that elements 107 and 109 could be produced by cold fusion reactions (the production yield could be estimated) and that the isotopes ^{262}107 and ^{266}109 decay by alpha emission. However, a direct identification was not possible and one important value, the half-life, could not be measured. The solution of this problem was found by the scientists in Darmstadt where a new heavy ion laboratory came into use at the end of the 1970s.

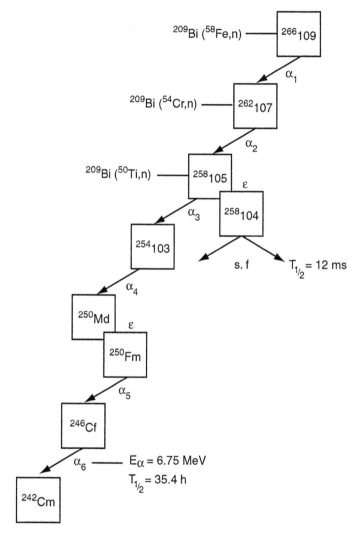

Figure 6.7 Sequential α and electron capture (EC or ε) decay chain for doubly-odd nuclei $^{258}105$, $^{262}107$ and $^{266}109$, leading to long lived ^{246}Cf production. The reactions for the synthesis of these nuclides are also given. The figure was taken from Oganessian *et al.* (1984).

7

The dream: uranium beams

Super and super-superheavy elements

Uranium plus uranium: that would be the heaviest possible combination of terrestrial elements. Imagine – element 184 with an atomic weight of 476! Would they fuse? What would the nucleus look like: a sphere, a disk or more like a bubble? What could be the lifetime of the product? How would it decay? What density could a material made of element 184 have? Would it be a liquid or a metal? How would the electrons arrange themselves around a nucleus with charge of +184? How would it react chemically? Questions galore!

Back in reality, progress in heavy element research was slow in the mid-1960s. The breeding of elements in reactors or nuclear bomb explosions ended at element 100. Relatively small cyclotrons could accelerate only 'light' ions. However, theory had made considerable progress. Vilen Mitrofanovich Strutinski, working in Kiev, developed a method of calculating the stability of heavy elements. The news spread like wildfire: across the world, and especially at the main centres in Copenhagen, Frankfurt, Berkeley and Los Alamos, theoreticians worked out the stability of heavy, superheavy and super-superheavy elements.

Magic nuclei and double magic nuclei were the centre of interest. But perhaps unlike the law of gravity, the theory describing the nucleus cannot simply be extrapolated over a wide range. It includes parameters which are fitted to known data. How well would it fare for describing nuclei which are far, very far, from the known region?

Theoreticians using a range of model parameters obtained different results for the prediction of magic numbers. For the protons it was 114, 120, 126 in the region of superheavies, and 154 and 164 for the super-superheavies. For the neutrons the magic numbers were 184 and 196, and 228 and 308, for the super and super-superheavies, respectively (Figure 7.1). These predictions encouraged experimentalists to fantasise while uncertainty fuelled speculation (Figure 7.2).

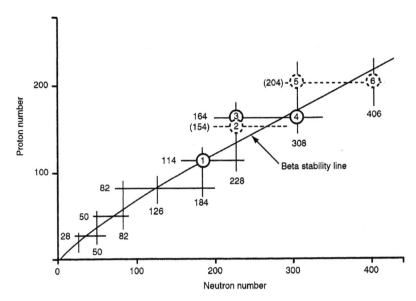

Figure 7.1 Calculated location of the islands of superheavy elements, 1, and super-superheavy elements, 2–6, in the nuclear chart. The figure was taken from Sobiczewski (1974).

The vacuum at high electric fields

Although the synthesis of superheavy elements was not the only dream of nuclear physicists in the mid-1960s, it was probably the most important. Another was quite extraordinary: developed in Frankfurt by Walter Greiner and his group of theoreticians, their dream was the study of a vacuum at high electric fields. Strange things were expected to happen. The project became famous as 'Electron–positron pair creation'. It is related to superheavy elements in two ways:

1. the very highest electric fields can be obtained in collisions of uranium nuclei. Even if the nuclei will not fuse to form a super-superheavy of atomic number 184, there will at least be a transient field for a short time when the two nuclei lie very close together.
2. the second aspect was a matter of getting beam time for experiments. That sounds a bit cryptic but all will become clear in due course.

The Frankfurt group calculated that, if two bare uranium nuclei devoid of all electrons were brought very close together, the binding energy for an electron in the innermost orbit would be 1.33 MeV. That created excitement – we will soon see why – and they pondered more deeply.

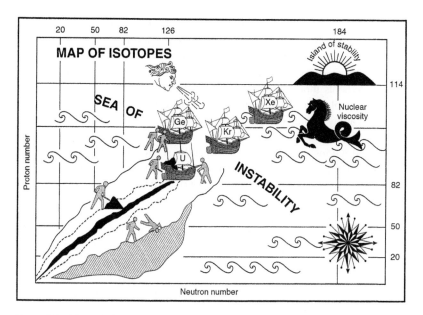

Figure 7.2 Allegorical presentation of the journey to superheavy island according to G.N. Flerov (1974). Ships mark various accelerator enterprises to reach superheavy island.

Remember Einstein's equation, $E = mc^2$? In particle physics only pairs can be created, pairs of particle and antiparticle, never a single particle alone. Bearing in mind that energy corresponds to the mass of an electron and its antiparticle, we find that the creation of an electron–positron pair requires a minimum energy input of 1.02 MeV. Let us compare these two numbers, 1.33 MeV binding energy for an electron and 1.02 MeV electron–positron pair creation energy. The binding energy is more than enough to create an electron–positron pair. Is it conceivable that one could *simultaneously* create an electron–positron pair, trap the real electron in an inner orbit of the combined uranium–uranium system, use the binding energy to create the pair and emit a real positron? A fantastic idea but well worth exploring. The upshot was that for the electron–positron pair creation, and for the synthesis of superheavy elements, an accelerator for beams of elements as heavy as uranium was necessary!

The UNILAC

Uranium is the heaviest element in nature. Because ^{238}U (which constitutes 99.3% of natural uranium) has a very long half-life of 4.5 billion years, it is in effect only weakly radioactive and can safely be handled and controlled in large quantities. So an accelerator for uranium poses no special safety problems and can provide the heaviest possible element as a beam for use in experiments. However, it was not just uranium that was of interest; the lighter elements thorium, bismuth, lead, gold, samarium, xenon and others were also to play a part. Particularly for the investigations of fusion reactions, a whole variety of projectile-target combinations was necessary. And bearing in mind that a heavy ion laboratory would be equipped with highly valuable and expensive instruments – not to forget the scientists – there should be no limitation for using even lighter elements like nickel, iron, calcium, argon, neon, carbon, helium and hydrogen. Indeed, all the elements needed potentially to be available to help to answer a variety of questions: a 'universal' accelerator was needed.

The plans were made in Heidelberg. Christoph Schmelzer and his small team of accelerator experts racked their brains about what such an accelerator might look like. It had to have high currents and a certain minimum beam energy so that uranium nuclei in the beam, directed on uranium targets, would be able to overcome electric repulsion (each uranium nucleus carries a charge of +92 at a very high charge density) and impact the target nuclei.

Soon it became clear that the accelerator they had in mind could not be a cyclotron (Figure 4.1). With any reasonable machine dimensions, only the extremely high charge states of ions emitted from an ion source could be used. However, the intensity of ions emitted from an ion source decreases rapidly with increasing charge state so the intensity of highly charged ions would be too low for practical experimental work. Intensity is at its maximum when only a few electrons are removed. In addition, it is difficult to deflect high intensive beams out from the cyclotron. The people in Dubna already knew about these difficulties from the work with their own cyclotrons some years previously.

A ring accelerator, a so-called synchrotron was a possibility. However, synchrotrons need time from the injection of a packet of particles to the attainment of the energy ultimately required and this reduced beam intensity. Furthermore, the beam would have a pulsed structure which would be unfavourable.

The solution could be a linear accelerator: what about building a direct-current voltage accelerator? At the time these were the work horses for accelerating light nuclei in many nuclear physics laboratories. The basic idea was developed by Robert Jemison Van de Graaff. He used a charging rubber belt to create voltages of up to 5 million volts, enough to accelerate protons and other light nuclei to velocities sufficient to result in nuclear reactions.

Heavier nuclei could be accelerated using a brilliant trick which resulted in the construction of a two-stage tandem accelerator. That trick was to start the acceleration with negatively charged ions, those which carry additional electrons. These are accelerated in the first half of the tandem from ground potential to the plus 5 Megavolt terminal in the middle of the tandem. There, the ions have already reached a high speed so that electrons can be stripped off when transiting a stream of gas or a thin foil. After stripping, the ions are positively charged and are now further accelerated from plus 5 Megavolt to ground potential giving a total accelerating voltage of 10 Megavolts.

Limitations include the fact that not all elements allow for producing negatively charged ions and that, even with modern high voltage generators, the direct-current voltage is limited to about 15 Megavolts. This is sufficient to accelerate ions up to mass unit of about 60 but not up to uranium of mass unit 238, even if negatively charged uranium ions could be produced; the uranium nucleus is simply too heavy due to the number of its neutrons.

Was this the end of a dream to accelerate uranium? It was not. If high enough voltages cannot be reached in one step, why not add many steps each of lower voltage? That is indeed a good idea but unfortunately does not work with a number of small direct-current voltage accelerators. When leaving the first accelerator the ion would be slowed down before entering the second because the voltage always changes between some positive high voltage and ground potential.

The solution was to use high frequency electric fields for acceleration rather than a direct-current voltage. It was the ingenious idea of Christoph Schmelzer to combine various types of high frequency structures which in total allowed for accelerating ions as heavy as uranium. A UNIversal Linear ACcelerator (UNILAC) was the result of his efforts.

The UNILAC is a high frequency (HF) accelerator

Attempts to use high frequency electric fields for the acceleration of particles go back to the inventions of Rolf Wideröe in 1928 and of Lawrence and David. H. Sloan in 1930. The principle of the method is shown in Figure 7.3. The even and odd electrodes are connected to opposite poles of an oscillator. In the gaps between the drift tubes located on the axis of the cavity there is an electric-potential difference. An ion in the gap is subject to the related field. An ion travelling inside the drift tube does not sense any field, being neither accelerated nor decelerated. If an ion crosses the gaps at the appropriate times, it receives multiple accelerating impulses. The distance between the gaps, i.e. the length of the drift tubes, has to increase if the ion is to cross the gap at the right time at increasing speed.

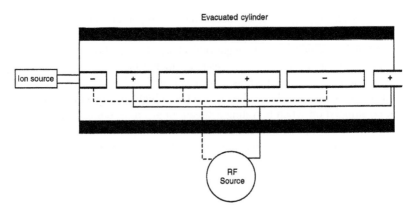

Figure 7.3 High frequency linear accelerator structure according to Wideröe (1928). The figure was taken from *History of Linacs* at http://www.accsys.com/history_linacs.html.

Acceleration starts with positively charged ions from an ion source; all accelerators require an ion source. For nuclei, many types of ion sources have been developed with continual improvements and new designs. The aim of all these efforts is further to increase the number of ions (i.e. the current) to be extracted and to increase the charge state of the ions. Figure 7.4 shows the principle of an ion source in which electrons coming from a hot filament oscillate in a volume occupied by a gaseous material and ionise it by collision. A whole distribution of various charge states is produced by most of them, unfortunately, at a relatively small value. In the case of uranium the maximum intensity is at about charge state 6^+ (this is so because uranium, 92, has six electrons outside the closed electron shell filled at radon, 86). The ions are pulled out by an electric field generating ion currents of the order of milliamperes.

The positive ions are first accelerated by a relatively low direct-current voltage of 320 kV to an initial velocity, the jumping-off point for further acceleration in the high frequency section of the UNILAC. The whole ion source must be at this positive high voltage because the main part of the accelerator has to be on ground potential.

Before entering the high frequency part of the UNILAC, the direct current of ions is formed into bunches of a few nanoseconds (billionth of a second) duration. These bunches have a repetition time of 37 nanoseconds which meshes with the high frequency of the accelerator. This first part is of the Wideröe type, shown in Figure 7.3. A view into the interior of the Wideröe section is shown in Figure 7.5. The bunches are accelerated

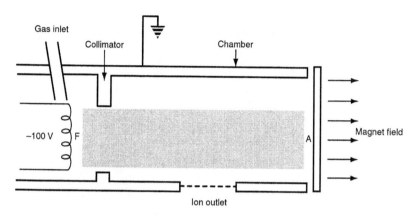

Figure 7.4 A typical positive-ion source. Electrons emitted by the filament *F* at −100 Volts oscillate between it and the anode *A*, while they curl around the lines of force of the containing magnetic field. The electrons form positive ions by impact in the gas present in the chamber. The mixture of positive ions and electrons, called plasma, fills the shaded region. Positive ions are extracted from the plasma through the aperture. The figure was taken from Segrè (1965).

between the drift tubes seen in the figure. They enter the field-free interior of the tubes when the accelerating voltage outside has changed sign and thus is decelerating the particles.

The Wideröe has a length of 28 metres. It accelerates the ions up to an energy of 1.4 MeV/nucleon which is still too low for nuclear reactions. However, the energy is high enough so that the charge state can be efficiently increased by stripping. This is done using carbon foils or a gas jet (nitrogen), the latter, although less efficient in stripping, is required in the case of high beam currents. After stripping, uranium ions of charge state 28⁺ are used for further acceleration.

The stripping of electrons was the first trick which Schmelzer's team used for the acceleration of uranium. The second was to increase the high frequency for the acceleration after the Wideröe section and stripping by a factor of four. Instead of 27 MHz (1 Mega-Hertz = 1 million cycles per second), a frequency of 108 MHz was used in the following sections. In this way the diameter of the resonator tube could be four times smaller which considerably reduced the volume of the whole accelerator. The volume itself was not really the problem but, with the reduced size, the costs of construction, vacuum, buildings and operation were all significantly reduced. The electromagnetic field inside the 108 MHz section oscillates so that an

Figure 7.5 Just like links in a chain: the 130 accelerating electrodes of the Wideröe structure are neatly arranged one after the other in the four copper-plated steel tanks. When the ions emerge from the metal cylinders, they are subjected to the force of a high-voltage field which accelerates them until they enter the next cylinder.

electric field along the axis arises. And as in the case of the Wideröe section the ions are not affected inside the drift tubes when the field between them has the wrong sign. This principle was developed by Luis Walter Alvarez and this part of the UNILAC was named the *Alvarez section*.

There are four such resonators, each 13 m long. They accelerate the ions in rather large steps from 1.4 to 11.8 MeV/nucleon. In order to obtain accurately the energies between the steps given by the Alvarez sections, fifteen single cavity resonators are installed behind the Alvarez sections. For heavy element synthesis we need energies of about 5 MeV/nucleon. In that case only Alvarez sections one and two are used and, with few single resonators, the beam energy can be accurately adjusted to any desired value around 5 MeV/nucleon. This is a very important feature of the UNILAC because the yield for the production of heavy elements is strongly dependent on the beam energy chosen.

I want to add one more peculiarity of the UNILAC which is a disadvantage for many experiments but turned out to be very useful in heavy element experiments. This characteristic is caused by the electric power which the high frequency radio tubes must provide. The power consumption is so high that these tubes cannot run continuously and deliver a quasi-direct current beam of particles. They are operated at a yield of about 30%; this gives rise to a macrostructure of the beam of 6 millisecond-wide macropulses followed by beam pauses of 14 milliseconds. The period during the beam pauses is completely free from projectile background, a positive benefit for making measurements.

Figure 7.6 is an overview of the whole accelerator complex as it looks now. The main part is the 120 metre-long linear accelerator UNILAC. To the left are the two ion sources; PIG stands for *Penning Ionisation Gauge*, the special type of ion source in use. Since 1993 another ion source (an Electron Cyclotron Resonance source) and a small Radio Frequency accelerator are in operation which allow for accelerating ions up to 1.4 MeV/nucleon and injecting them directly into the Alvarez section. *SHIP*, the main instrument for heavy element research, is indicated. It is located in the low energy experimental hall where beams from 2–17 MeV/nucleon are available. For experiments at high beam energies, a synchrotron accelerator was built at the end of the 1980s: a synchrotron is a special kind of ring accelerator and SIS stands for *Schwer Ionen Synchrotron* (Heavy Ion Synchrotron). It uses the UNILAC as injector and accelerates the ions up to an energy of 2 GeV/nucleon (Giga-electron Volt, 1 GeV = 1000 MeV). This energy corresponds to a speed greater than 90% that of light.

The construction of the UNILAC and research using heavy ion beams became the mission of a new research centre, the *Gesellschaft für Schwerionenforschung* (GSI). Founded in 1969, it is financed and maintained 90% by the Federal Republic of Germany and 10% by the State of Hesse; an aerial view of the GSI is shown in Figure 7.7. The first heavy ion beam was generated at the end of 1975.

First experiments and disappointments
The Nuclear Chemistry Group led by Günter Herrmann at the Radiochemistry Institute of the University in Mainz, already well-established in new element and transuranium element research, was the one best prepared to

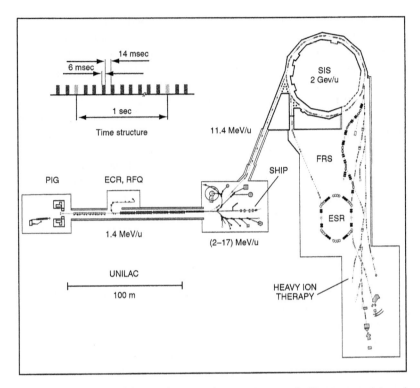

Figure 7.6 Plan view of the accelerator and experimentation facility. Ions are injected by the two ion sources in the left-side named PIG (Penning Ionisation Gauge). They pass through the 120-metre-long UNILAC which consists of three sections: 1, the Wideröe structure; 2, the Alvarez structure; 3, a total of 15 single resonators for a fine tuning of the beam energy. The maximum speed of the ion beam after the UNILAC is 16% of the speed of light. The greatest part of the beam is directed to the adjoining experiment hall for experiments, among those of the SHIP for the synthesis of superheavy elements. A second ion source and injector were ready for experiments in 1993. The set-up consists of a special ion source named ECR (Electron Cyclotron Resonance source) and an accelerator structure named RFQ (Radio Frequency Quadrupole structure). Only a small fraction of the beam which can be delivered from one or the other of the two injectors is directed into the heavy ion synchrotron SIS for further acceleration. There, the ions reach up to 90% of the speed of light before being diverted to experiments at the fragment separator FRS, the experimental storage ring ESR or further down into an experimental area which also houses the medical laboratory for tumour therapy. The inset at the upper left of the diagram shows the time structure of the beam: six millisecond wide pulses followed by a beam pause of 14 milliseconds. Only once per second is one of the pulses deflected into the SIS.

Figure 7.7 Aerial view of the GSI research centre upon completion of the second phase of expansion. The large buildings on the right house the heavy ion synchrotron, the experimental storage ring, the related detectors and, since 1997, a medical laboratory for tumour therapy. The SHIP is located in the hall with the dark roof in the centre of the photograph while the UNILAC accelerator extends inside the tunnel at the lower left.

start the search for superheavy elements. Herrmann had been preceded by Straßmann who, in the years following 1935, worked with Otto Hahn and Lise Meitner at the Kaiser Wilhelm Institute for Chemistry in Berlin. Searching for new transuranium elements in 1938, it was that team which discovered nuclear fission as described in Chapter 1.

One of the very first reactions was Ar + Pb and the detection system was a reproduction of one used in the Russian instrument at Dubna. From the very start, the biggest problem was the low melting point of lead and the target was soon destroyed even by the modest beam currents then in use.

The aim was to study uranium plus uranium interactions, not really with any expectation of observing a super-superheavy element 184 – that may have just been a secret dream. The objective was to search for transfer products, reaction products obtained when part of the projectile material is transferred to the target nucleus or vice versa. This process is shown schematically in Figure 7.8.

A pair of interacting uranium nuclei has a neutron to proton ratio of $146/92 = 1.59$. This is almost identical to that for the projected superheavy nucleus $^{298}114$: $184/114 = 1.61$. Might it not be possible for the target or projectile nucleus to capture from its reaction partner exactly the right numbers of neutrons and protons so that the smallest of the theoretical superheavies $(Z = 114)$ is formed? This would require the transfer of 22 protons and 38 neutrons, a nucleus of the element titanium $(Z = 22, A = 60)$. Such a nucleus

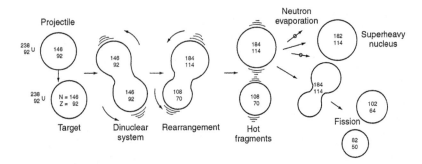

Figure 7.8 Scheme for the synthesis of superheavy nuclei by transfer of nucleons. Two colliding uranium nuclei (actually not spherical but deformed similar to eggs) stick together for a short time. It was hoped that strong shell effects at proton number 114 and neutron number 184 would create a force and sufficient protons and neutrons to be sucked in from the reaction partner to fill up the shells. The resulting heavy nucleus is excited and will usually fission but might survive in a few cases after cooling down by emission of neutrons. However, this view was not confirmed by experiments. The figure was taken from Herrmann (1988).

(^{60}Ti) really exists; it is one of the most neutron-rich titanium isotopes which was discovered recently among fission products.

It was thought that there might exist some driving force especially for this transfer, that uranium would try to fill up its empty orbits of nucleons and form the double magic superheavy, just as in chemistry the elements try to fill the empty shells of electrons and form a noble gas. But something must have been wrong with these suppositions. Although the radio chemists searched very carefully for traces of superheavy elements, the heaviest element they could observe was mendelevium, element 101, even in the more favourite reaction $^{238}U + ^{248}Cm$.

The experiments were really hot, radiochemically hot (see Box 7.1), not just exciting. In order to cover a whole range of energies, the experimenters used relatively high beam energies and thick targets. In the target the projectiles are eventually degraded down to energies far below the interaction barrier when they no longer can participate in any nuclear reactions. However, all the way down from their initial high energy state to the ultimate low levels, the projectiles do react with the transfer of various small numbers of protons, neutrons, α- and other light particles. There is also a good deal of fission of both projectile and target nuclides. As you can imagine, everything

---

Box 7.1 The meaning of 'hot'

Every human activity, in every language, throws up its own jargon. Nuclear chemistry is no exception.

'Hot' simply means very radioactive: a lot of radioactive decay is taking place in a 'hot sample'. Of course, as in other fields, a topic can be hot and so can a flask left on a heater. Somehow nobody doing this sort of work gets these meanings confused!

---

gets extremely radioactive. Procedures had to be designed to handle such hot material and separate it carefully into the various fractions. A mainly automatic apparatus was built for this purpose.

Yields were measured and plotted in the form of cross-sections for each element detected and for each isotope of those elements. Large cross-sections were measured for isotopes close to the projectile and target nucleus but they decreased rapidly for the heavier elements. The value measured for the production of mendelevium (101) was 100,000 times less than that for berkelium (97), the element with the highest cross-section produced by transfer of only one proton from uranium to curium (96).

What could have been wrong? Why did the superheavies not show up? Later on, when we discuss reaction mechanisms, we might find a possible answer. Meanwhile the search for superheavies in transfer reactions was abandoned worldwide; everyone concentrated on fusion reactions.

8

Models, cross-section and fusion

Something about models

Models have an important role in life. Writing this sentence, I am of course thinking about beautiful models wearing beautiful dresses but, as we know, models do not represent reality in all respects. So it is with nuclear models. The reality contained in the detail is much too complicated for us to deal with the whole picture when we aim to investigate only one particular aspect. Take a house, for example: city planners need just a simple wooden model of the right size for their planning mock-ups –the interior is of no interest. Architects and heating engineers, on the other hand, need to have models which show the heat capacity and conductivity of the walls. Models come in various sorts depending on what one needs them for.

The simplest nuclear model is that of a solid charged sphere with a certain mass and diameter. That is quite enough to describe correctly the elastic scattering of α-particles as was done by Rutherford. If we are going to describe the nuclear binding energy and some aspects of fission and fusion, the charged liquid drop model is a good approximation, one we have already discussed in some detail. We get the same results if we consider the nucleus as a cloud of gas with the nucleons as gas particles; such a view of the nucleus of gold is shown in Figure 8.1. There is enough space between the nucleons for them to move freely with a velocity as much as 25% that of light and that is what nucleons (protons and neutrons) inside a nucleus actually do. The natural consequence of so much rushing about is the diffuse nuclear surface but the diffuse surface is a property difficult to understand with the liquid drop model. The idea that the nucleons move around randomly is a very good approximation (model) for a highly excited (or hot) compound nucleus which results from a fusion reaction.

We can refine our gas model so that protons and neutrons do not move around randomly but in well defined orbits like the electrons in the electron cloud around the nucleus; then we end up with the *nuclear shell model*, our most detailed model for the nucleus. This explains correctly the number of protons and neutrons needed to fill each shell, the magic

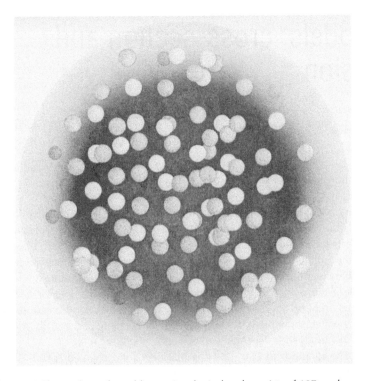

Figure 8.1 The nucleus of a gold atom is spherical and consists of 197 nucleons: 79 protons (lighter stippling) and 118 neutrons (heavier stippling). Although the nucleus is 10^{15} times denser than the remainder of the atom, even in it there is a great deal of 'empty space'. Each nucleon has a diameter of approximately 1 femtometre (10^{-15} metre); the centre-to-centre distance between neighbouring nucleons is some 2 femtometres. The density of nucleons and the strength of the intense nuclear force (dark grey) are uniform in the interior of the nucleus and fade off gradually toward the surface (light grey). The nuclear force does not drop off abruptly with the last nucleon but diminishes slowly. Here and there in the surface of the nucleus, two protons and two neutrons coalesce into an α-particle. These are not immutable structures; the clusters continually appear and dissolve in the nuclear surface as the individual neutrons and protons from the interior shuttle in and out of the surface as they follow their separate orbits. Inside the nucleus, protons and neutrons are moving with velocities as high as 25% of that of light. The figure was taken from *Scientific American*, October 1972.

numbers, reproduces correctly the structure of the binding energy across the whole chart of nuclei and allows for the prediction of the stability and location of nuclei close to the next double magic shell closure, the superheavy elements. In addition, the model shows us which orbits are filled for the ground-state of a nucleus, the angular momentum and the structure of the excited energy levels above the ground-state. It also predicts the shape of a nucleus: spherical, oblate like a pancake, prolate like a rugby ball, or more like a pear, all of them possible nuclear shapes.

What is a 'cross-section'?
Nuclei interact in various ways. What actually happens when a target is irradiated with projectiles depends on the properties of the interacting nuclei: their sizes, shapes, masses and charge numbers, the energy of the projectile and how far off centre is the collision, the *impact parameter* of the reaction.

The projectile and target nuclei actually very rarely approach one another closely enough to interact, as rarely as hitting the jackpot in a lottery. A major factor is the very small size of a nucleus. Think of the sort of lead target actually used in our experiments, $1\,cm^2$ in area, 0.4 micrometres (1/25,000th of a centimetre) thick and with a weight of 450 micrograms (less than 1/2,000th of a gram). Such a target contains 1.3×10^{18} atoms, each consisting of 82 protons, 126 neutrons and 82 electrons.

If we assume that all nuclei are all in a plane (not an unreasonable assumption because it is extremely rare that two nuclei are exactly one behind the other), from the projectiles point of view all the 1.3×10^{18} atoms cover an area of $2.1 \times 10^{-6}\,cm^2$ out of the total area of the target of $1\,cm^2$– two parts in a million – so the chance of collision is pretty small. This follows from the lead's nuclear radius of $7.1 \times 10^{-13}\,cm$ ($=7.1$ fm, femtometres). Now suppose we start to irradiate this target with ^{70}Zn projectiles. The zinc nucleus has a radius of 4.9 femtometres. If the energy is high enough it will always hit the target nucleus whenever the impact parameter is less than the sum of the two radii, i.e. with 12.0 femtometres the interaction area increases from the earlier $2.1 \times 10^{-6}\,cm^2$ to $5.9 \times 10^{-6}\,cm^2$. Divide this area by that of the target (still $1\,cm^2$) and the chance of hitting just one target nucleus in a one-shot experiment is still no better than 1 in 170,000. It really does remain a lottery.

The term 'target nucleus' is somewhat misleading! What we really can hit with the beam is the $1\,cm^2$ target, not the nucleus. The nuclei are hit by chance with the probability we have just calculated. Now suppose there were just one target nucleus for each square centimetre of the lead target instead of the 1.3×10^{18} which are actually present: we would divide the actual nuclear target area (these were the $5.9 \times 10^{-6}\,cm^2$) by 1.3×10^{18} to get the extremely small number $4.5 \times 10^{-24}\,cm^2$ – and the chance of hitting one lead nucleus with the zinc projectile is 1 in 222,000,000,000,000,000,000,000. That is about equal to the probability of hitting the eye of a fly sitting somewhere on the earth with a grain of sand of similar size thrown from the moon. However, it would not be so bad if we were aiming not at one fly but at 0.65×10^{18} of them (which gives us 1.3×10^{18} eyes to aim at because each fly has two!). Then the total area of their eyes would be $13,000\,km^2$. It sounds odd, doesn't it, but physicists sometimes have strange ways of trying to make clear their ideas to non-physicists!

The area $10^{-24}\,cm^2$ has a special name in nuclear physics: 1 *barn,* abbreviated 1 b. It is the dimension of a quantity which we call the *cross-section* of a reaction. In our example, the cross-sectional area of a lead nucleus waiting to be hit with a high energy zinc nucleus has a value of 4.5 barn. Divided by $1\,cm^2$ it gives the probability for such an event and we are back to those big numbers. This value of 4.5 barn comes only from the dimensions of the projectile and target nuclei; it tells us nothing about what happens if they collide.

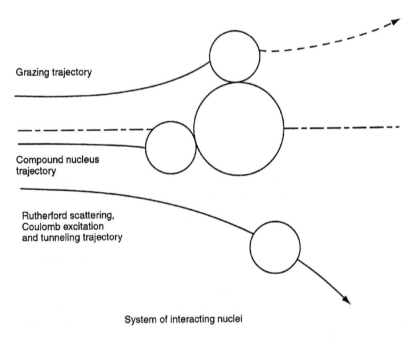

Grazing trajectory

Compound nucleus
trajectory

Rutherford scattering,
Coulomb excitation
and tunneling trajectory

System of interacting nuclei

Figure 8.2 Diagram showing the various possibilities for how nuclei can interact. Please, keep in mind that nuclei look as shown in Figure 8.1, they do not have a sharp surface as shown here in this simplified drawing.

The various ways in which nuclei can react are sketched in Figure 8.2, which shows the *grazing trajectory* which was the basis for our cross-section example. This trajectory is curved upwards due to the electric (Coulomb) repulsion between projectile and target nucleus. However, the curve is modified when nuclear forces start to act at the surface: the trajectory is so far out that the projectile is not caught but continues its flight more or less undamaged. The cross-section is strongly energy-dependent: if the energy decreases, the projectile nuclei are more easily deflected away from the target and the cross-section for a nuclear reaction decreases.

Central collisions most likely result in the *fusion* of projectile and target nucleus but this does not mean that the fusion product, the compound nucleus, actually survives without fissioning. For the synthesis of superheavy elements the *fusion cross-section* has to be multiplied by the *survival probability* in order to get the total production cross-section. The survival probability is a very small number in the region of the heavy elements and the total cross-sections approach values as low as 1 picobarn, this is 10^{-12} barn.

Finally, the projectile may be lucky and pass the nucleus without direct collision although, even at a distance, it will be affected by the Coulombic repulsion and the more closely it approaches the target nucleus, the more it will be deflected. This is 'elastic scattering' because the only change is to speed and direction of motion and not to the projectile and target nucleus themselves. It was this type of collision which Rutherford observed, from which he concluded that the atom consists of a tiny, but heavy nucleus surrounded by electrons; in honour of Rutherford, the phenomenon is also called 'Rutherford scattering'. The cross-section for Rutherford scattering is much higher than for nuclear reactions because the projectile is scattered at a distance and does not have to hit the target nucleus.

But not all cases of elastic collisions are ideal. Sometimes kinetic energy from the relative motion of projectile and target is transferred via the *electromagnetic* field into excitation of the nuclei. I emphasise *magnetic,* because moving charged particles are surrounded by a magnetic field in addition to the electric one. One impressive mode of this Coulomb excitation which happens to deformed, rugby ball shaped nuclei, is rotation around the short axis. When these rotating nuclei come to rest, a very regular pattern of gamma rays is emitted; from their energy we can deduce the shape of the nucleus, a very elegant way of looking into the microcosm of nuclei.

The fusion reaction
Physicists try to simplify physical phenomena, necessary in order to understand the essentials and to quantify the observations in physical laws. Phe-

nomena can be separated into static and dynamic categories. A nucleus in its ground-state is an example of a static object – perhaps 'quasi-static' would be more accurate because the protons and neutrons inside the nucleus are moving with velocities as high as 25% of that of light. Nevertheless, the nucleus will not change its properties: it is a stable system – or quasi-stable if it is radioactive and hence stable only for a limited period, its lifetime.

Dynamic processes are more complicated to describe. Everything is moving and appearances change with time; kinetic energy is converted back and forth into potential energy, angular momentum is distributed among particles and removed by radiation, order is replaced by chaos and order later reappears.

The fusion process of two nuclei is the example which interests us most. For we can reduce the complex dynamic process into a sequence of quasi-static pieces; those steps are shown in Figure 8.3. We will take as an example the fusion of ^{70}Zn and ^{208}Pb which resulted in ^{277}112.

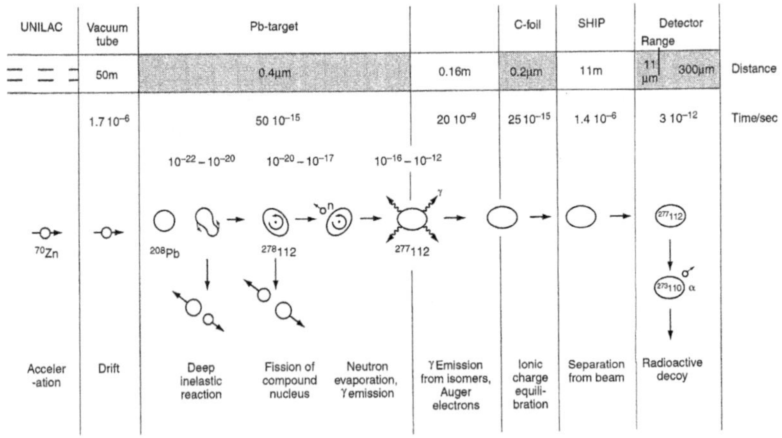

Figure 8.3 The various stages in the formation of nuclei $(Z > 100)$ in heavy-ion collisions and their subsequent decay. The example shows the formation of element 112. Only a very small fraction of the projectiles (1 in 100,000) collides with a target nucleus. Again, only a small fraction results in the creation of a rotating but quasi-stable compound nucleus. Most of those 'hot' compound nuclei break into two pieces by fission, especially if the rotation speed is high. In our example of the element 112 synthesis, only one compound nucleus per week cooled down by emission of one neutron and subsequent γ-rays. This evaporation residue was separated from the beam by SHIP and implanted into a detector, where it decayed by α-emission. The various stages of the reaction are characterised by a timescale and the location of where the various phases occur.

The first step is the approaching phase. The projectile leaves the accelerator with the target nucleus at rest. The projectile becomes progressively more influenced by the electric field of the target nucleus as it crosses the electron shells. There is, of course, interaction between the electrons of both the projectile and of the target but the energy exchanged is on a level of electron-volts (eV) for the outer shells and some thousands of eVs (keV) for the inner shells. This is negligible compared with the tens or hundreds of MeV involved in the nuclear interaction.

The second step is probably the most complicated and dynamic in a nuclear reaction and ought to be divided into a number of sub-steps. It begins with the first contact of projectile and target nucleus and ends (hopefully) with the formation of a compound nucleus, the united system. The projectile's kinetic energy is converted partly into potential energy, partly into rotational energy and partly conserved in the movement of the combined system. This is the kinetic energy of the centre of gravity formed by the projectile and the target nucleus which has a constant value during the whole reaction process.

If the impact parameter (the deviation of the projectile's path from a central collision) is too large, the centrifugal forces within the system of the two nuclei which have just touched will be too high, only few nucleons exchange in either direction before the projectile and target nucleus separate again. At a smaller impact parameter, the nuclei remain together longer and more nucleons can be *transferred* – the term normally used. The repulsive Coulomb and centrifugal forces are nevertheless still too large and the system will separate but, this time, after exchanging more nucleons; indeed, so many can exchange that the process is difficult to distinguish from the fission of the compound nucleus formed later. But the angular distribution, the direction into which the fragments are emitted, is different and this enables two processes to be separated experimentally. In the first case we speak of a *quasi-elastic* collision, in the second of a *deep inelastic* collision; some people also call it *quasi-fission* or *fast fission*. Only when the collision is almost head on, resulting in small angular momentum and low centrifugal forces, will the projectile be trapped and a compound nucleus formed.

We have to keep in mind that the ratio in which the nuclei will be scattered quasi-elastically, deep-inelastically or captured depends strongly on the projectile energy and on the mass and charge number of the reacting nuclei. At low charge number, let say $O + Sn$ ($Z = 8 + 50$), the probability of capture is high even at large impact parameter. For heavy element synthesis, as in our example $Zn + Pb$ ($Z = 30 + 82$), the capture probability is extremely small and the other processes predominate.

After capture, the compound nucleus forms, the system comprising the sum of the protons and neutrons. If the impact parameter was not zero, it rotates and, in our example, has an *excitation energy* of 10 MeV (equivalent to temperature) which originates from the differences in binding energy (remember the exothermic chemical reactions) and the surplus of kinetic energy. The more kinetic energy the projectile has, the higher it can heat up the compound nucleus. This, however, is a disadvantage for heavy element synthesis because hot compound nuclei strongly tend to fission. We prefer the neutrons to evaporate; they are bound in the nucleus by about 8 MeV so, in our example, at 10 MeV excitation energy, only one neutron can be emitted; that is the same as saying that one neutron is enough to cool the compound nucleus down to an excitation energy somewhere between 0 and 2 MeV, the exact value determined by the amount of kinetic energy of the emitted neutron.

A decisive factor for the synthesis of heavy elements is the ratio between neutron emission and fission. Unfortunately, this is very small and has a decreasing trend with increasing element number. It can be also considered as a ratio between partial lifetimes of the compound nucleus for fission and neutron evaporation. If the fission lifetime is short, fission predominates. How long the lifetimes actually are for fission and neutron evaporation of a hot compound nucleus is one of the unanswered questions but values in the range of 10^{-17} to 10^{-20} seconds are thought likely.

If a neutron is emitted, we are almost home. Neutrons are emitted most favourably just straight away from the nucleus. Then they carry no angular momentum so the nucleus is left rotating with an excitation energy between 0 and 2 MeV. The rotational energy and the rest of the excitation energy is removed by the emission of γ-rays. This takes place within 10^{-16} to 10^{-12} seconds and the nucleus then reaches its *ground-state*. As it was produced by cooling due to the evaporation of neutrons, we term the final fusion product an *evaporation residue*.

We already know the time spans for the various processes and from them it is possible to determine where exactly the processes took place. The projectile has a velocity of 10% of light, i.e. 30,000 km/sec (or 3 cm/nsec; 1 nanosecond $= 10^{-9}$ seconds). The path length from the UNILAC to the target is 50 m so the travel is 1.7 microseconds. It takes a projectile 13 femtoseconds (1 fsec $= 10^{-15}$ sec) to traverse the 0.4 micrometre thick target, a time so short that it is difficult to imagine but for the projectile it is a period full of activity. During that time it meets exactly 1,276 lead atoms (if it does not collide with a target nucleus). It is then easy to calculate how long the projectile needs to travel half the atomic diameter from the outer electrons to the lead nucleus. It is just necessary to divide 13 fsec by 1,276 and then by 2 to get 5.1×10^{-18} seconds. Actually the time is somewhat longer because the projectile is slowed down as it approaches the nucleus. Let us assume that the rest of the kinetic energy of the projectile relative to the target nucleus is 10 MeV when it touches the lead nucleus, corresponding to a velocity of 0.9% of light or 27 femtometres/10^{-20} sec. The radii of the lead and ^{70}Zn nuclei are 7.1 and 4.9 femtometres, respectively, leading to the conclusion that the ^{70}Zn nucleus needs about 1×10^{-20} seconds to travel from one side of the lead nucleus to the other, more or less along a trajectory at the surface. This is the contact time, a relatively long period for a nucleus.

Once more, a lot happens to the nuclei during these 10^{-20} seconds because protons and neutrons inside the nuclei are moving at 25% the velocity of light or 7.5 fm in 10^{-22} seconds. Thus, during the 10^{-20} seconds contact time, some of the protons or neutrons can easily travel from the projectile to the target nucleus and back a hundred times, time enough to 'decide' where they would like to remain.

The compound nucleus has a velocity of 2.6% of that of light. It needs four times longer than the projectile (50×10^{-15} seconds) to travel through the full target thickness (if it was created just at the beginning of the target). So all processes which are shorter than that take place inside the target. In Figure 8.3 we see that this period covers the emission of most of the γ-rays. It means that the newly created nucleus has already reached the ground state, complete and in its final shape, ready to start life when it leaves the target. But it is no time to rest: the first excitement after being born is separation from the many projectiles which passed by the target without reacting.

Although we are now running far ahead of our story, I would like, using Figure 8.3, to continue for a moment to follow the complete lifecycle of our newly born ^{277}112. As an example of a separator I will use our SHIP instrument to be described in the next chapter. In the SHIP, the distance from the target to the detector is 11 metres, which governs the time for separation of the fusion products from the beam at 1.4 microseconds. If our nucleus survives less than that, it will decay somewhere inside the SHIP and most probably even the daughter nucleus will then be lost.

After separation, our ^{277}112 reaches the detector. Like a meteor crashing to earth, although very much faster, the ion crashes into the detector surface at a velocity of 2.6% of that of light. The detector is made of silicon and its electrons splash away at the impact; sometimes even whole silicon atoms are removed from their site in the lattice. This costs energy and the ion comes to rest after 3×10^{-12} seconds, in that time travelling 11 micrometres into the silicon. There it fills up its empty electron orbits and tries to find a space for itself between the silicon atoms. If nothing disturbs it further, it can already now begin to enjoy a well-earned retirement after a dramatic birth, exciting childhood and turbulent youth.

The death of [277]112 follows after 280 microseconds, accompanied by the simultaneous birth of [273]110. The daughter element is born by the emission of an α-particle, the birth itself taking only 10^{-22} seconds. The α-particle is emitted with a velocity of 8% of that of light, coming to rest after travelling 90 micrometres through the silicon. Thus, using 300 micrometre thick detectors, we can be sure that all the α-particles come to rest inside the detector, except those which are emitted in the reverse direction and have to travel only 11 micrometres before they escape from the detector.

The newly-born daughter experiences only a small shock at birth resulting from the recoil of the α-particle, giving a recoil velocity so small that [273]110 comes to rest within 0.1 micrometre, 37 lattice sites away from its place of birth. A new lifecycle begins and will end again with the emission of an α-particle. However, that story comes later.

9

New techniques – kinematic separators and position-sensitive detectors

In previous chapters we have seen how fusion products were studied using mechanical devices like rotating wheels and cylinders. In all of them the fusion products are stopped and transported to detectors. That takes time. The fastest method was the Russian cylinder which could measure half-lives down into the range of milliseconds but was sensitive only for fission fragments. For the measurement of α-decay one must place the detectors further away from the target because the energy of α-particles is about 10 times smaller than the energies of fission fragments, so more sensitive detectors must be used. These, however, would suffer from too high a radiation background near the target. To overcome this problem, tape devices were developed which moved the activity a few metres away. Gas transport systems (helium jets) transported the activity in a gas stream (usually helium) through a thin capillary several metres long. With these methods, the detectors could be placed far away from the target in a well-shielded area, although the transport time increased to several seconds so that all nuclei with half-lives from microseconds to seconds would be lost. A completely new separation technique had to be developed and this became possible with kinematic separators.

Kinematics is the science of motion and kinematic separators make use of the movement of the reaction products together with another property inevitably connected with motion, the charge state of the ion – the number of electrons stripped off from an atom when it moves through matter. Removing electrons means that the charge of the nuclear protons predominates and the ion is positively charged. The number of stripped electrons depends on the ion's velocity; the faster it is going, the more electrons are stripped.

What does a separator for fusion products do? Fusion products are only a small fraction of the total reaction products: in the case of superheavy element formation, an unbelievably small fraction. However, they do have a well defined energy and direction of travel which is determined by the laws of the conservation of energy and momentum (see Box 9.1).

It seems that we have here ideal conditions for the separation of the

Box 9.1 The conservation of energy and momentum
The classical laws governing energy and momentum have been known
for a long time:

- *energy can neither be created nor destroyed* so the total energy
  within a system remains constant;
- *a moving body has a momentum defined as its mass times its
  velocity*. That momentum is maintained as long as our body (the
  projectile) is not exposed to forces acting in gravitational, electric,
  magnetic or other fields and dramatically in a collision with
  another body. If that body (the target) is at rest, then its momentum
  is zero because its velocity is zero. After the collision the two of
  them (the compound nucleus) will travel with a *total momentum*
  given by the total mass times a new velocity *in the direction* of the
  moving particle. This total momentum, however, is equal to the
  momentum of the projectile, because the target momentum was
  zero. Therefore we can easily calculate the new velocity. It is the
  projectile velocity times projectile mass (which is the projectile's
  momentum) divided by the total mass (mass of projectile plus mass
  of target). Because the total mass is always bigger than the projectile mass, it follows that the velocity of the compound nucleus is
  always smaller than the velocity of the projectile. It is much
  smaller in the case of light projectiles directed on heavy targets,
  only slightly smaller in the opposite case and exactly half the projectile velocity in symmetric reactions.

fusion products from all other reaction products which leave the target with
all kind of energies in all kind of directions. However, the main problem
remains, the separation of the fusion products from the beam particles which
pass through the target without reacting. They leave the target travelling forwards, like the fusion products, but at much higher velocity so that they are
distinguishable.

Unfortunately, there are some disadvantages. The velocities and directions of the fusion products are averaged out by the recoil momentum resulting from the emission of one or more neutrons from the compound nucleus
and also by collisions with other target atoms. Moreover, the electric charge
state of the particles is not well defined. The beam velocity is 10% of light;
for the reaction $^{58}Fe + ^{208}Pb \rightarrow ^{265}Hs + 1n$ (as an example), that of the fusion
products is about 2%. When leaving the target, the fast beam particles have,

on average, lost 22 electrons out of 26 (iron is element 26) but more or fewer electron losses are possible. We end up with a charge distribution which follows almost a normal curve with half-width of four charge units.

The problem is that normal curves have tails, long tails, so that extremely low and extremely high charge states also occur. This would not be serious at low beam intensities but at high currents these extremes create problems for the separator, because ions with low charge state are less deflected in electric or magnetic fields. They are described as being 'stiff' or as having high electric or magnetic 'rigidity' when they are not greatly deflected even in strong magnetic or electric fields because of their low charge state, high mass or high velocity. For example, at a beam intensity of 6×10^{12} particles/sec, the flux of ^{58}Fe ions (in our Fe + Pb reaction) with charge states less than +10 is still 10^5 particles/sec. Preventing as many as possible of these stiff ions from reaching the detectors is a further important task of a separator.

The separator's second task is to allow passage of the fusion products and direct them to the detector system. But the fusion products also have a charge distribution. In our example, the ion ^{265}Hs of element 108 (after one neutron evaporation and having left the target) has an *average* charge state of 24 but this statement hides the fact that the actual charge on any particular ion might be anywhere from about 21 to 27, with small probability of values outside those limits.

Such situations are typical and we have to compromise: how many 'good' particles we can afford to lose and how many 'bad' ones we can accommodate. That is a most difficult decision for the designers of a separator (especially for a prototype) when all these numbers are merely theoretical.

When it became clearer that the UNILAC would be built at Darmstadt, the scientists there began to define an efficient separator for superheavy elements. It should be:

- based on physical principles;
- fast, faster than all previous technology and thus complementary to the chemical methods; and
- optimised specifically for fusion products.

Three types were considered:

1. a *gas-filled separator* (in the original proposal it was actually a velocity filter combined with a gas-filled separator) proposed by Peter Armbruster and co-workers at the Nuclear Research Centre at Jülich;
2. a *velocity filter* of the Wien type proposed by Heinz Ewald and co-workers at the University at Giessen;

3. a *high frequency filter* proposed by Egbert Kankeleit and co-workers at the University at Darmstadt.

The GSI research centre chose the second for the experiments at the UNILAC: it is known as the *velocity separator SHIP* for reasons we will soon learn. But other groups chose differently: the accelerator team at the University in Munich chose the third while groups in Berkeley, Dubna, Tokyo, Jyväskylä (in Finland) and later also at GSI (as a second separator for testing purposes) used the first. Before we explore the workings of SHIP, let us have a look at the other separators; that may explain why the choice is not an easy one.

Gas-filled separators
The gas-filled separator contains helium or hydrogen at low pressure (approximately 1 torr). Ions moving through the gas can be considered to carry only one charge state of certain mean value. The reason is that ions exchange electrons in collisions with the atoms of the gas. When they lost too many electrons, the electric forces increased and, in the subsequent collisions, electrons were recaptured from the gas atoms or molecules. Averaged over many collisions, the ions lose a well-defined *mean* number of electrons. Because electrons are lost, the charge of the nucleus predominates and the ion is positively charged. The gas pressure is set so that the distance between the collisions is short compared with the path length of the ion in a magnetic field. Only then can the mean charge state be used as the relevant number for the deflection of the ion in the magnetic field. However, the gas pressure must be low enough for the ions not to lose too much energy in too many collisions on their path through the separator; they still have to reach the detector at the far end.

The separation of the fusion products from the beam particles is achieved by a magnetic field; fusion products and beam particles are subject to different deflections by this magnetic field called *different magnetic rigidities*. There is a simple relation for the deflection angle in a magnetic field as function of the mass, the charge state and velocity of a particle. The deflection angle becomes larger when the charge state increases or when the mass and velocity decrease. For many reactions, the beam particles are deflected more than the fusion products, shown in the drawing of the Dubna gas-filled separator in Figure 9.1. The beam is stopped at a cooled metallic plate, whereas the fusion products can pass through the gap of the magnet.

Behind the deflecting magnet are two magnetic lenses, so-called quadrupole lenses, which focus the fusion products to the detector. The magnetic fields of two quadrupole lenses act on charged particles like a magnifying glass. Why focusing is necessary will be explained a little later when dealing with the SHIP.

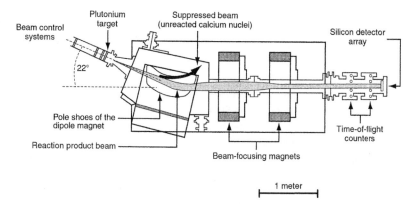

Figure 9.1 Scheme of the Dubna gas-filled separator GNS which was used for the study of the reaction $^{48}Ca + ^{244}Pu \rightarrow ^{292}114^*$ and $^{48}Ca + ^{248}Cm \rightarrow ^{296}116^*$. The figure was taken from *Scientific American*, January 2000.

Gas-filled separators work well for highly asymmetric reactions as in the use of light-weight projectiles against a uranium target; in such cases, few beam particles reach the detector. With heavier projectiles, especially in symmetric reactions, the magnetic rigidity of the fusion products leads to increasing number of projectiles as unwanted background on the detectors; this effect limits the use of gas-filled separators.

High frequency filters

The high frequency filters, to which we referred earlier, make use of the fact that at high frequency the particles are accelerated and hit the target in bunches. In the UNILAC the bunches last for 1–2 nanoseconds and occur 37 nanoseconds apart. This time structure is determined by the first section of the UNILAC high frequency accelerator, working at a frequency of 27 MHz which results in an oscillation period of 37 ns.

Kankeleit's idea was to install another high frequency resonator behind the target, working at the same 27 MHz frequency (or an integral multiple of it) but with an electric field transverse to the beam direction. At a certain distance behind the target the oscillating electric field would kick the beam bunches in one direction and the slower bunch of fusion products in the other. It was a brilliant idea and relatively cheap to put into effect. However, to find the optimum distance behind the target, the resonator must be movable and its position individually adjustable for the various reactions. Another disadvantage is that there would be a high frequency electric field close to detectors and sensitive electronic devices with a consequent risk of electrical noise seriously disturbing signal processing. In the event, the Kankeleit proposal was rejected and Ewald's group at the University in Giessen was awarded the project.

The velocity separator SHIP

Professor Ewald was *the* expert on mass spectrometers in Germany. In Figure 9.2 we see him, the 'father' of the SHIP, with Professor Christoph Schmelzer, the 'father' of the UNILAC. At the time, Ewald had two young physicists in his group, Klaus Güttner and Gottfried Münzenberg. The three of them were working on a proposal for an electromagnetic velocity filter. The final design of a two-stage filter originated with an idea of Münzenberg's. They were joined later by Wolfgang Faust (a graduate student), the main driving force at GSI for the installation of the SHIP. His doctoral work was the first of the experiments done with SIS, the *Schwer Ionen Separator* (heavy ion separator), SHIP's original name.

Figure 9.2 Photo showing Heinz Ewald (left), the father of SHIP, and Christoph Schmelzer, the father of UNILAC, at a party celebrating 10 years of SHIP operation in March 1986.

At GSI, Peter Armbruster had, since 1971, been scientific leader of the divisions of Atomic Physics and Nuclear Chemistry II (the German is *Kern-Chemie II* [KCII] which actually was the SHIP group). Possibly inspired by Flerov's allegorical picture of the conquest of the superheavy island (Figure 7.2), Armbruster baptised the apparatus 'SHIP', the *Separator for Heavy Ion reaction Products*.

I was recruited to the KCII/SHIP group in October 1974, immediately after my doctoral thesis on gamma spectroscopy in Kankeleit's institute. A year later we were joined by Karl-Heinz Schmidt and Willi Reisdorf; an electronic engineer Hans-Joachim Schött and technician Hans-Georg Burkhard completed the group. Since then the whole activity has grown enormously and we now have scientists visiting from all over the world.

A schematic view of the SHIP is shown in Figure 9.3; this is the modified version in operation since 1994. In the original, the target was further away from the first magnetic lenses and the final 7.5 degree deflection magnet did not exist.

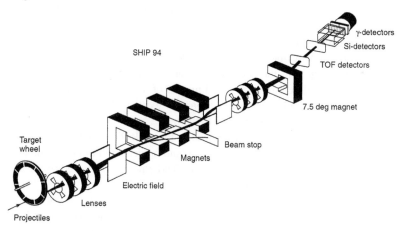

Figure 9.3 The key instrument for the discoveries of the elements 107 to 112, the velocity filter SHIP at GSI Darmstadt. The upgraded 1994 version is shown. The 11-metre-long instrument separates fusion products from the projectile beam during a flight time of 1–2 microseconds. It consists of two electric and four magnetic dipole fields, two quadrupole triplets and a final deflection magnet. The drawing is roughly to scale with the target wheel and the detectors enlarged by a factor of two. The target wheel of about 35 cm outer diameter rotates synchronously with the time structure of the beam. The detector system consists of two secondary electron time-of-flight detectors and a position-sensitive silicon detector array. New elements are identified by proton and neutron number through their α-decay into long decay chains consisting of known nuclei.

The SHIP is basically a velocity filter, deriving from an idea by the physicist Wilhelm Wien as far back as 1897–8; velocity filters were at first called *Wien filters*. Wien used it for the analysis of charged low energy heavy ions created in the gas of an electric discharge; he found that in crossed magnetic and electric fields, all the particles move with the same velocity, independently from their mass or charge, in a circular orbit of fixed radius.

The Giessen scientists very soon realised that an ideal velocity filter could not be built for heavy ions of high velocity like the fusion products of a heavy ion reaction; their dream, remember, was still that U + U might fuse to form super-superheavies which a newly designed instrument should, of course, be able to separate. To overcome the technical difficulties, they simply separated the electric from the magnetic field. Then, instead of a full circle (which would have brought the fusion products back to the target), they used an electric field giving a deflection angle of only −6°. The electric rigidity of the beam particles is much higher than their magnetic counterpart, passing through the electric field almost undeflected. In Figure 9.3 the direction of the deflection is to the left looking downstream, indicated by a minus sign.

The first of the following magnets bends the fusion products back by +12°, with the next one set again for a deflection of −6° which brings the particles back again to the zero degree axis. This combination compensates for different electric charge states and masses of the particles in the medium plane. In the end, only a velocity dispersion exists with the faster particles located more to the right, the slower to the left.

The choice of 6° is a compromise. If the deflection angle were larger, the fusion products with higher charges would be deflected too far to the left, running out of control in the decreasing field strength at the edges of the magnets or stopped in the walls of the vacuum chamber.

Quantitatively, the following relations are important: the deflection angle in the electric field is ~ $q/(m\ v^2)$ and in the magnetic field ~ $q/(m\ v)$. We see that the velocity dependence of the location of particles in the medium plane comes into the game due to the $1/v^2$ dependence of the deflection angle in the electric field and the $1/v$ dependence in the magnetic field. If a smaller deflection angle were chosen, the beam would not be deflected far enough to the right and a larger fraction of beam particles would pass the SHIP and could reach the detector.

Although the beam was deflected sufficiently by the first three fields, it was nevertheless decided to add a second, point symmetric stage. This means the deflection angle in the third magnet was +6°, in the fourth magnet −12° and in the second electric field again +6° degrees. Two considerations were responsible for this decision.

1. Projectiles and other reaction products scattered through the velocity slit in the medium plane would have a fair chance to be filtered out in a second stage. As a result the background would be considerably smaller in the focal plane, where the detectors are located.
2. The several hundreds of thousands of volts needed in a heavy-ion velocity filter are difficult to measure. A voltage error in only one field would direct the fusion products in the wrong direction. Therefore, the plates of the second electric field are supplied from the same high-voltage generators for positive and negative voltage as the first one.

We have seen how the beam is separated from the fusion products by using electric and magnetic fields. We have finally to deal with the problem of letting as many fusion reaction products as possible pass through the separator. Although a fusion reaction looks very simple and straightforward, some properties make this difficult. First are the various number of charge states distributed around a mean value. But that is not enough. The fusion products also have different velocities. Imagine the reaction happening at the front surface of the target with the ion product travelling through the whole target thickness, colliding with target atoms, each time losing a small amount of velocity before it escapes from the target into SHIP. On the other hand, if the reaction were to take place near the exit surface of the target, the fusion product would immediately escape into SHIP without losing velocity. The target thickness thus defines two extremes of velocity (and all values in between occur) but only a certain range of velocities is acceptable to SHIP: the slow fusion products from the reactions at the beginning of target, in particular, are likely deflected too much and hit the walls of the vacuum chamber. The useful target thickness for kinematic separators is therefore limited to values of about one micrometre.

It is a third effect which makes additional magnetic fields necessary, the so-called quadrupole lenses. The reason is that the fusion products, although having various charge states and velocities, do not travel in the exact beam direction. They get their main kick out of beam direction when the compound nucleus (i.e. the hot intermediate stage in a fusion reaction) cools down by emission of neutrons. These neutrons are ejected at high speed giving rise to a recoil velocity. Because the neutrons are emitted equally in all directions some residues receive a sideways kick, so changing direction, while others are kicked forwards or backwards, so changing the velocity of the fusion product. All combinations of these effects are found. In addition, as the fusion product travels through the target, it collides with target atoms, each time changing direction a little. Summing up, a considerable distribution of angles around the beam direction occurs.

Without magnetic lenses, most of these diverging reaction products would be lost.

At SHIP we use a combination of three lenses at the entrance and at the exit (see Figure 9.3). The lenses are magnetic quadrupoles having four magnetic poles, north and south, alternating every 90° around the circumference. The combination of three such lenses (the *quadrupole triplet*) acts for ions just like zoom optics in a camera. At SHIP the first triplet focuses the fusion products diverging from the target to the medium plane where the particles are separated by velocity. There, a 50 millimetre-wide aperture lets only the wanted fusion products pass. The second triplet focuses those particles onto the detector.

A sharp focus can be obtained only for ions all of similar velocity and one charge state. But the particles we want have a number of charge states and different velocities. That demands a compromise in the field gradient of the quadrupoles which determines the focal length. This is most difficult because the input data, the distribution for charge states, velocities and the angles are not known exactly. This compromise is nevertheless critical because it determines the number of fusion products which finally reaches the detector as well as the number lost. Calculations indicate a ratio (number of fusion products on the detector divided by fusion products created at the target) to be 1/2 for reactions with lead targets and ought to move closer to 1 as the SHIP is upgraded.

The arrangement, of the three main components of the SHIP, the quadrupole lenses, the plates for the electric field and the dipole magnets is shown in Figure 9.4. The whole of the SHIP interior is under high vacuum, 10^{-7} torr. The distance from the target to the detectors is another important factor as it determines the shortest measurable lifetime because the particles must reach the detector before they can be identified. The length of the SHIP is 11 m and the resulting flight time 1–2 microseconds depending on the mass and speed of the projectile.

Construction of the SHIP took from 1972 until 1974. The components were delivered from the manufacturers at the end of 1974 and a year later the first test experiment was run. Construction and installation were almost flawless: only a few minor defects showed up which were easily put right. Many of the smaller electrical and mechanical problems had to be solved on-line and our working day ended often at midnight. It was an exciting time and exciting, too, was the prospect of synthesising superheavy elements.

Figure 9.4 Photograph showing the main parts of the velocity filter SHIP. From the left: 1, two quadrupoles of the first quadrupole triplet; 2, two cubes for the creation of an electric dipole field. The cubes are charged up to voltages of about +100,000 Volts and −100,000 Volts, thus creating a strong electric field between the plates which are 15 cm apart. The vacuum vessel above the two cubes was removed; 3, two of the four dipole magnets. These are electric magnets and we can see the isolation of the copper coils surrounding the iron. A current of about 100 Amperes creates the necessary magnetic field strength.

The choice of detectors: first, the rejected devices

Running parallel with the installation of the SHIP was the development of the detectors, as difficult in itself as the design of the separator. There were many unknowns and people had many different opinions about how to do it.

First of all, we did not know how well SHIP would work and how many unwanted particles we would see at the detector position. The flux of this background would determine which kind of detectors could be used. At a background particle rate of 10^6 per second, as suggested by estimates, it would not be possible to use solid state detectors. These are made of silicon and are destroyed after being hit by too many heavy ions, with a total of 10^{10}

about the limit. That meant replacing the detector after three hours at a cost each time of about $1,000.

Second, we could not be sure how big an area on the detector would be impacted by the fusion products. The ion optical properties resulted in a 1:1 magnification which means that the image of the beam spot on the target is displayed on the detector at its original size. However, the ion optical elements (the magnets and the high voltage deflectors) were designed for high transmission, not high resolution. Their ion optical quality was too poor for producing a sharp 1:1 image. How large could the diameter be: 10, 20, 50 mm? Larger yet? Would small detectors be enough or would we need larger, more expensive ones?

Third, how would superheavies decay? Would we need fission or α- or β-detectors? Would it be wise to install neutron detectors because some predictions were that a fissioning superheavy would emit lots of neutrons? And what about γ or X-ray detectors? Or might there even be a strange decay mode so far unknown? Remember the electron–positron pair creation predicted for super-superheavy elements discussed in Chapter 7.

And finally, what half-lives should we expect? Microseconds or even less? Seconds, minutes, hours? Perhaps even years or hundreds of years; might we observe a long lived superheavy nucleus? It made a lot of difference: the longer the half-life the less activity we would observe per unit time.

Armbruster had the most information about superheavies; he was already on the conference circuits when the rest of us were still taking our degrees. He was very much in favour of methods which do not require the radioactive decay of the fusion products because he believed in almost stable superheavies with very long half-lives. In that case what are needed are detectors which allow for measurement of the element number and also, if possible, the mass number of the produced nuclei. He made two proposals.

1. To measure the X-rays which are emitted when fast ions pass through a thin foil. The energy of X-rays is characteristic for the element number of the ion. An X-ray is emitted when an electron of the ion moves from an outer orbit to an empty place in an inner orbit. He proposed using the transitions from the third orbit into the second and to generate an empty place in the second orbit by letting the ion pass through a thin tin foil. Although the energy of the transition from the second to the first orbit is higher and would offer more advantage, it is more difficult to remove an electron from the first orbit.
2. To apply a well known method of reaction physics for identifying iso-

topes: measuring the ions' velocity and energy allows for extracting the ions' mass. Armbruster suggested complementing the measurements by a device for determining the loss of energy in a thin foil, a loss characteristic for the element number of the traversing ion. As a result we would get what we wanted: the mass and the element number, sufficient for a complete identification of the fusion product and applicable for all ions independent of the nuclear half-life.

Both ideas were brilliant and simple but, like many brilliant and simple ideas, they didn't work. This does not mean that they didn't work at all: we tested the methods for lighter elements and concluded that they would not work for the superheavies. For the X-ray measurements, the speed of the ion would be too low for the production of a hole in the second orbit. This difficulty could be overcome by using the inverse reaction which means a heavy projectile combined with a light target. Due to the conservation of momentum, the speed of the fusion product would be much higher; however, the flux of background projectiles would be enormously increased, preventing the use of sensitive detectors. The same argument – too low a speed of the ions – holds for velocity, energy and energy loss measurement. At low speed the detector resolutions were too poor to reliably identify the fusion products.

We also thought about systems which would allow the detection of superheavies in the presence of a high flux of background projectiles. In that case the detectors could not be placed directly at the focal plane of SHIP and, in order to avoid their destruction, we had to resort to mechanical devices not so much for a long transport distance as for killing the unwanted background particles.

One such mechanical device which we actually tested was a tape system behind SHIP. The idea was to stop all particles, background and fusion products, in a movable tape. The non-radioactive background projectiles would then not harm the tape as it moved between sensitive detectors; only the radioactive nuclei would be measured. This method would, of course, not work for stable or long-lived superheavies because radioactive decay is the way they signal their existence.

However, tape units never worked properly, just gave us problems and needed incessant maintenance and improvement. Bearing this in mind, I looked around for tape systems used in industrial production where such problem devices are not acceptable. Finally, more or less by chance when speaking to a salesman of paper computer tapes, I found a factory near Stuttgart. Those people were producing pullovers with the design programmed on a perforated paper tape, actually a tear-resistant plastic material 0.1 mm thick and 25 mm wide. They used a tape, about 200 m long, made endless by gluing the ends together, necessary because the design covered the main circular part of the pullover. The tape ran at high speed through a capstan drive which stopped and started the tape; after going through the drive, the tape looped into a container from the bottom of which it was pulled out and run to the reading device from which the needles were operated. This was exactly what we needed (except for the pullover and the needles!); well tested, simple, fast and relatively cheap.

The idea was to put the tape behind the SHIP. Through a thin window, the reaction products emerge from the SHIP vacuum into air and are deposited onto the tape. After a certain collection time, adjusted to the half-life of the nuclides, the tape was moved forward 15 cm. The deposited activity would then be positioned exactly between two detectors: a silicon detector for α-particles on the deposition side (there the escaping α's lose less energy) and an X-ray detector behind the tape. This was an ideal geometry for the measurement of α-particle/X-ray coincidences, i.e. whether both emissions were detected at exactly the same time and hence probably came from a single source. The tape was quite fast with a transportation time for the 15 cm of about six milliseconds, making it possible to measure half-lives down to several milliseconds.

X-rays of specific energies are characteristic for particular elements while α-particle/X-ray coincidences provide a unique method for identifying an element. The X-ray energy can be calculated quite accurately for any new element, because the Bohr atomic model works so well. Tables already exist up to element 180 or so.

The other instrument we built was a helium jet transport system. Its purpose was to trap the reaction products in a gas volume and blow them via an extremely short capillary (5 mm) onto a collector or directly onto the surface of a silicon detector. Trapping needed a certain gas volume and it took time to wash this out; using the helium jet meant that the lowest measurable half-life was about 200 milliseconds. This again illustrates the limitations of mechanical devices. In the most favourable case the shortest measurable lifetime would be few milliseconds and we would therefore have lost a wide range of possible products with half-lives ranging from a few microseconds to milliseconds. That would have been a pity but luckily it soon turned out that the mechanical devices were superfluous.

The gold makers of Darmstadt
Tension mounted as the time approached for the first beam through the SHIP. It happened on December 18th, 1975.

At 8 o'clock that evening we generated a ^{129}Xe beam which was maintained for four hours. Only the first half of the SHIP was ready and we wanted to test the deflection of the low energy beam by the electric field, but for some reason we failed.

The following two attempts to produce and separate fusion products were a disaster. On February 20th, 1976, we planned to use another Xe beam but the evening before the high voltage failed. On February 25th we did get a ^{40}Ar beam lasting six hours. Eventually we realised that one of the magnets was connected wrongly to the power supply so the reaction products were not being focused. Newly-designed devices always seem to behave like this: whether they are particle physics instruments, the new model of a car or anything else electrical or mechanical, it invariably takes time to iron out the bugs. (You may remember the 'Elk Test' of a new model of a very prestigious brand of car some years ago.)

However, the next irradiation on March 12, 1976, was fully successful. We had a ^{40}Ar beam and a target of ^{144}Sm. We had mounted a silicon detector in the focal plane and behind it a germanium X-ray detector. The spectra looked marvellous – beautiful and with many lines showing good resolution (Figure 9.5). We could even watch the lines growing. After only a couple of minutes we were ready to stop data collection and try to understand what we had. Soon it was clear that we had done what the alchemists of old had only dreamed of: we had made gold (a pretty expensive way of doing so, I must admit!). And not only gold: there was mercury, platinum, iridium, osmium, rhenium and tungsten.

The background rate was low. The high energy part of the spectrum measured with the silicon detector is also shown in Figure 9.5. The main fraction in the spectrum was a broad bump at 51 MeV with a much smaller peak at 11 MeV, resulting from scattered argon projectiles having the same velocity as the reaction products and therefore passing through the velocity filter. Left of the argon peak were a number of very narrow lines which we identified as α-lines because they derive from α-particles. On the expanded part we could, by their energy, assign the lines to known isotopes. This told us about the reaction ^{40}Ar + ^{144}Sm → ^{184}Hg*. The asterisk means that this is the hot compound nucleus formed by the sum of protons and neutrons of the projectile and target. This compound nucleus cools down by the emission of neutrons and sometimes also of protons and α-particles. What remains are the evaporation residues separated by SHIP and directed onto the detector where they come to a halt near the detector surface. The detector is able to convert the kinetic energy of the particle stopped in the detector into an electric signal which is further amplified. Finally, the voltage maximum of the signal (usually 0–10 volts) is measured. After calibration against known standards, the precise energy of the particle is calculated from the height of the signal.

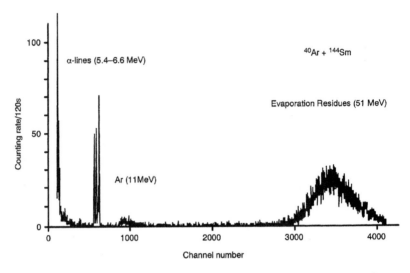

Figure 9.5 Typical energy spectrum taken with a silicon detector at the focus of SHIP. The spectrum was measured using a counting time of only two minutes during irradiation of a ^{144}Sm target with ^{40}Ar projectiles. The channel number is proportional to the energy, the energies of the structures being given in MeV in brackets. At 51 MeV we see the evaporation residues produced in the reaction, separated by SHIP and implanted into the detector. A small peak of argon projectiles which passed the SHIP is detected at 11 MeV. The α-lines from the decay of the evaporation residues and their daughter products are in the energy range from 5.4 to 6.6 MeV. To the left we see a steeply rising background which originates from electrons of β-decaying nuclei.

The α-particles emitted from the radioactive nuclei previously implanted in the detector were similarly measured with the silicon detector. Finally, we measured with the germanium detector the characteristic X-ray spectrum; its energy and structure is clear evidence that elements were produced. The lines we saw were the so-called Kα- and Kβ-X-rays; they are emitted when an electron moves from the second orbit into a empty place in the first orbit (Kα) or from the third into the first (Kβ). Here, the empty places are produced by the electron capture decay of the radioactive nuclei (see Chapter 1). Not shown is a spectrum displaying the region of highest energy where we would expect to find beam particles of the original energy (236 MeV) as projectile background; there were no such particles. SHIP had deflected all of them to the beam stop. There was good reason that evening to break out the champagne!

Production of superheavies – our first disappointment

After that successful run of gold and platinum production we eagerly awaited the attempt to make our first superheavy. It took place as Run 20 on December 3rd, 1976. We were really optimistic and thought it could be done in one day. But getting access to the UNILAC beam was not easy because so many experimenters wanted to use the beam to do other things.

Our first superheavy reaction was ^{136}Xe $+ {}^{170}$Er$\rightarrow {}^{306}$122*. According to the predictions of the American theoreticians, Edmund O. Fiset and James Rayford Nix, we expected an α-decay chain from element 122 down to 114 with short half-lives: microseconds for the heavier isotopes increasing up to seconds for the lighter. Element 114 itself might undergo fission, α-decay or electron capture with a long lifetime of a few hours.

Within twelve hours we collected a considerable beam dose and found nothing. The second attempt ran from June 11th–13th, 1977; we were lucky to get two days of beam time. The intended reaction, ^{65}Cu $+ {}^{238}$U$\rightarrow {}^{303}$121*, was asymmetric resulting in higher excitation energy of the compound nucleus and more fission. However, the fusion cross-section was expected to be higher than in the case of the more symmetric Xe + Er. We collected product for 39 hours and again drew a blank. We had to conclude that the synthesis of superheavy elements 121 and 122 must have very small cross-sections or half-lives outside the range of one microsecond to one day, the limits for which we were set up.

I no longer remember in detail our great disappointment – disappointments are easy to repress – but the records do not refer to any bottles of champagne.

The learning phase

Fortunately nobody was pushing us too hard to search for superheavies so for a while we stopped and entered a learning phase. That is not yet over, and never will be, but the very important initial phase ended in January 1980. What did we have to learn? What did we actually learn? We had a new SHIP but we couldn't sail it. We wanted to reach the superheavy island but our voyage ended after two days!

The first thing was how to measure rare events. I emphasise 'measure'. We approached the detection of rare events in different ways; eventually we could combine the most efficient methods.

We were aiming for synthesising new isotopes. α-emitters were relatively easy to identify; they are marked in yellow on the chart of the nuclides (Figure 1.5). Karl-Heinz Schmidt and his co-workers started to explore the closed shell nuclei at $N = 126$, moving up through the elements thorium, protactinium, uranium. They discovered new isotopes and isomers, measured half-lives, α-energies and excitation functions – the yield of isotopes as function of the beam energy. They compared yields using asymmetric or symmetric reactions and in the latter found significant shifts for the beam energy to higher values in order to fuse the nuclei. Qualitatively these shifts were in agreement with theoretical expectations.

A highlight of his work was the discovery in 1982 of *radiative capture*. In the symmetric reaction of two ^{90}Zr nuclei (element 40), the group found complete fusion, unambiguously identifying the nucleus ^{180}Hg (element 80). This showed that nuclei can fuse without emission of protons or neutrons; they just emit gamma rays to cool down into the ground-state. That might be an important point for future superheavy element research and a good point here to refer to the 'fusion reactions' discussed in the previous chapter.

Willi Reisdorf was searching for effects on the cross-section below the barrier. At beam energies below the barrier ('sub-barrier energies') the cross-sections rapidly become small, and sensitive methods are needed to detect product. In reactions of a ^{58}Ni beam with targets of various samarium isotopes at beam energies below the barrier, he found that the cross-section was strongly dependent on the shape of the target nuclides. The light samarium isotopes are spherical while the heavy ones are deformed, like rugby balls: the more deformed the target nuclei, the higher the cross-section below the barrier. He also found a difference between nuclei which are spherical and stiff (like those close to double magic nuclei) and those further away from the double magics which are less stiff and tend to vibrate.

My own immediate area of concern was the neutron deficient isotopes near the magic neutron number $N = 82$. These are α-emitters with half-lives ranging from milliseconds to seconds, just right for measurements in the SHIP.

In July 1977 we were able for one day to use a beam of ^{58}Ni. For some reason we irradiated a ^{107}Ag target and were surprised by the many α-lines which rapidly grew in the spectra. The compound nucleus is ^{165}Re, element 75; at the time no α-emitting isotopes were known either of rhenium or of tantalum, the daughter element after α-decay. In our irradiation we immediately found 11 new isotopes of elements from hafnium to rhenium. It was a lucky chance that our first 'new' element (107) which was discovered later belongs chemically to the same group as rhenium and was sometimes called *eka-rhenium* ('beyond rhenium').

There are two reasons why these isotopes were not measured before; both emphasise again the advantages of kinematic separators. The isotopes have short half-lives of less than one second which hindered their observation with helium jets. Furthermore, these elements have high melting points and are not volatile; that makes them inaccessible in *on-line mass separators*.

These are different from kinematic separators like SHIP: in the former, the fusion products are stopped in heated (up to 2,000 °C) material whence they escape by diffusion to the surface. This costs time and differs from element to element depending on volatility; non-volatile elements do not escape at all. Thus, on-line mass separators were not generally discussed as instruments for finding superheavy elements; for instance, eka-rhenium (element 107) would be impossible to measure with an on-line mass separator.

In April 1981, the study of nuclei near $N = 82$ resulted in a new era of decay spectroscopy, *proton radioactivity.* Just as fission limits the production and stability of heavy elements, so does radioactive emission of protons on the neutron deficient side in the region of light and medium heavy elements. When very heavy nuclei carry too many protons they tend to fission but lighter elements emit α-particles until, at increasing proton numbers (i.e. moving upwards or to the left in Figure 1.5), they emit protons. The ability of a nucleus to emit protons is determined by their binding. In stable nuclei, the binding energy for the protons is positive and the protons cannot be emitted because they are bound energetically (remember the law of conservation of energy). For very neutron-deficient isotopes, electric forces repel the protons occupying the outer nuclear orbits so much that they finally become *unbound.* The proton binding energy becomes negative because the daughter nucleus and proton are more strongly bound when separate than being together in the parent nucleus. So proton emission becomes energetically possible.

Both, α- and proton-emission bring the daughter nuclei back closer to stability. This is evident from the curved (inclined to the right) location of the stable nuclei in Figure 1.5. After α-decay the daughter nucleus has two protons and two neutrons less and is located closer to the stability line than is the parent. Similarly, after proton emission, the daughter nucleus is exactly below the parent nucleus. On the very neutron-rich side (on the right side in Figure 1.5), the nuclei carry too many neutrons, so many indeed that finally one is no longer bound by the nucleus and is emitted. After neutron decay the daughter nucleus is located just left of the parent and, as in the case of proton emission, closer to the stability line.

An energy spectrum of the first case of proton emission from the ground state is shown in Figure 9.6. The line originates from the isotope ^{151}Lu (lutetium, element 71) with a half-life of 85 milliseconds.

The excitation function (the yield plotted as function of the beam energy or excitation energy of the compound nucleus) is also interesting. Figure 9.7 shows that, in the proton-unbound region, protons are not only emitted from the ground state but even more are emitted from the hot compound nucleus of a fusion reaction. As a result, it is difficult to obtain nuclei more neutron-deficient than those compound nuclei because the protons are

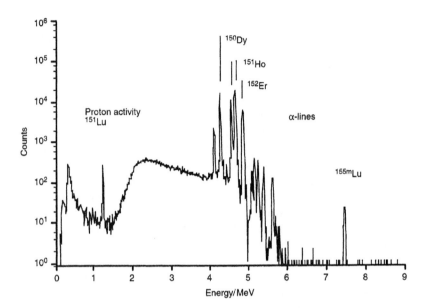

Figure 9.6 Energy spectrum taken from the reaction of a ^{58}Ni beam and a ^{96}Ru target. The compound nucleus ^{154}Hf is extremely neutron-deficient. By evaporation of one proton and two neutrons, the evaporation residue ^{151}Lu is produced. Together with other reaction products this nucleus is separated by SHIP. After implantation into a detector, its radioactive decay was measured: the new line, later identified as a proton line, was observed in the spectrum far below the α-lines originating from other reaction products. This was the first observation of a proton radioactive nucleus located beyond the proton drip line. The broad bump at 2–3 MeV energy is due to α-particles which escape backwards from the detector and so give up only a fraction of their energy. The α lines at energies above 5 MeV were produced from reactions with target impurities of heavier ruthenium isotopes. The figure was taken from Hofmann *et al.* (1982).

so weakly bound producing not lighter isotopes but lighter elements. In the chart of nuclei (Figure 1.5) this is a movement downwards instead of sideways and to the left. Fission plays a similar role in heavy element synthesis: losses due to fission of the compound nuclei increase tremendously and progressively reduce the yield for the production of heavy elements.

In Figure 9.7 we see that the cross-section for the synthesis of the proton emitter ^{151}Lu has a value of 100 microbarn at the optimum beam energy, which results in few hundred counts in the proton line shown in Figure 9.6. Assuming that the superheavies could be produced at cross-sections of 1 nanobarn or 1 picobarn, then the production yield will be one

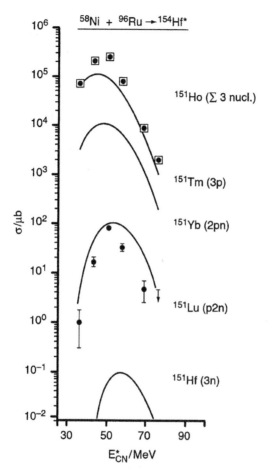

Figure 9.7 Comparison of experimental (symbols) and calculated (full curves) excitation functions of three nucleon evaporation channels of the reaction $^{58}\text{Ni} + {}^{96}\text{Ru} \rightarrow {}^{154}\text{Hf}^*$. The ^{151}Ho data are from a long lived daughter product and represent the sum of three nucleon channels. No data were measured for ^{151}Yb and ^{151}Hf. The abscissa gives the excitation energy of the compound nucleus which is related to the beam energy in a simple way. The curves exhibit maximum cross-sections at about 50–60 MeV and a strong cross-section decrease by six orders of magnitude from the 3 proton channel to the 3 neutron channel. The decrease reflects the low binding of protons in this region of nuclei which are much more readily emitted than are neutrons. In the region of heavy elements we have to replace proton emission by fission but then, of course, those nuclei are lost and only the pure neutron evaporation channels result in production of heavy elements.

hundred thousand or one hundred million times smaller. The relatively easy experiments at high yield opened our eyes for what we had to expect on the journey to superheavy elements.

Arrays of position-sensitive silicon detectors

The region of short lived α-emitters near $N = 82$ produced in high cross-sections was extremely useful for testing and developing various types of detectors as well as for investigating the ion optical properties of the SHIP.

We started with silicon detectors 24 mm in diameter. They had good resolution for α-particle spectroscopy and were also sensitive to fission fragments. Their thickness of 0.3 mm was enough to stop even the most energetic decay α's. Energy resolution could be improved considerably by cooling the detectors to about $-15°$, thus reducing thermal noise. Maintaining a high vacuum was a major factor in all our construction, influencing the materials we used for seals and other equipment.

Access to facilities was usually difficult; nuclear research facilities are very busy places, with lots of competition for access. At that time we usually analysed our data at night because during the day it was hard to find a free terminal and, when many people were working, our main central computer was very slow. Things was somewhat better during the night and there were no meetings or phone calls.

One night, around 2 a.m., I was analysing the α-spectra of the Ni + Ag run. Some of the data were collected by the position sensitive detector and I was watching tiny light flashes on the screen as they accumulated in a scatter plot of intensity distribution. It was easy to modify the programme in order for a period to follow what was happening at the point where the fusion product hit the detector. It was overwhelming. I found whole chains of α's: $^{163}Re \rightarrow ^{159}Ta \rightarrow ^{155}Lu$ or $^{162}Re \rightarrow ^{158}Ta \rightarrow ^{154}Yb$. Only the expected β-decay of ^{154}Lu was absent. Figure 9.8 gives the principle of the method and some of the data are shown in Figures 9.9. Clear as a bell, I got the half-lives of many nuclei in one shot, the data revealing the proportions of α- versus β-decay. I drove home very happily through the early morning.

What then had to be done became clear. I designed a detector array tailored specifically to the requirements of the SHIP. It consisted of seven position-sensitive detectors arranged vertically; the active area was $87\,mm \times 27\,mm$, interrupted only by the frames of the individual detectors. The specification went out to tender; it was delivered on September 18th, 1978, and I opened the box as if it were a Christmas present: inside the piece looked very beautiful, the golden frame, connectors in silver, the silicon shiny in dark grey glued into the white ceramic frames. I was over-whelmed.

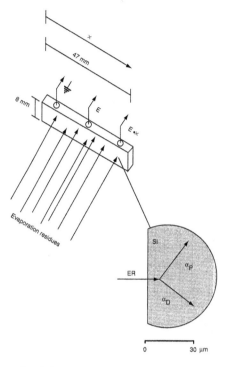

Figure 9.8 The principle of the position–time correlation method using a standard position–sensitive silicon detector 8 mm × 47 mm and 0.3 mm thick. The surface was specially treated for measuring the energy of particles as well as their position along the detector. The accuracy of the position measurement is about 0.8% of the detector length, i.e. for our detector 0.4 mm full width at half maximum of the position distribution. Assuming that the detector is homogeneously irradiated by evaporation residues with a counting rate of one per second, the interval between subsequent hits of the same detector position within twice the detector resolution is then one second × (47 mm/0.8 mm) ≈ 60 seconds. For measurement of lifetimes this means prolongation by a factor of 60 compared with the measurement without using the position information. The insert shows an enlarged cross-section of the detector near the surface where a radioactive evaporation residue was implanted. The implanted nucleus came to rest at a depth of about 10 micrometres, fixed at a location within the silicon crystal lattice. Depending on its lifetime, the nucleus will emit an α-particle ($α_p$) which has a range of about 30 micrometres at an energy of 6 MeV. The range of the recoiled daughter nucleus is much less than 1 micrometre: it moves only a few lattice sites from its original location, comes to rest and may also decay by α-emission ($α_D$). For the α-particles the same position is measured as for the implanted residue and, if the α-decay chain occurs within 60 seconds, it is uniquely identified as a correlated sequence of signals.

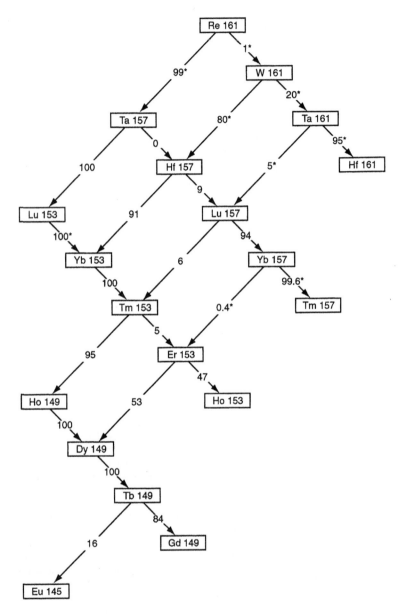

Figure 9.9 Decay pattern of ^{161}Re with α- and β-branchings in percentages. Long arrows represent measured α-transitions, the shorter arrows β$^+$-decays. However, the β$^+$-decays were not directly observed and the branchings deduced from the measured α-decay intensities.

Next I began to arrange the electronics for signal processing. Each one of the seven detectors needed two preamplifiers, followed by higher gain amplifiers for the α-decays and lower gain for the fission events. The signals provided us with the necessary information about the energy and position of the decaying nuclei. Additional electronics were needed to measure the exact moment when the signal was generated in the detector; time measurement was important for determining the lifetime of our product and eventually our accuracy reached one millionth of a second. All this kept me busy until February 1979 when we ran the first experiment using the new detector array.

Improving the prerequisites

A superheavy element experiment is very exciting, much more so, of course, for those actually doing the work than for external observers. It is like watching a race compared with being the sprinter yourself with all the training, balanced diet, choice of running shoes and the rest – to say nothing of the uncertainty.

Superheavy element experiments are a bit like that with lots of effort before the experiment even begins, much more than during the run itself. Only time will tell whether synthesis is possible; our task is to make the event visible, reducing, or better avoiding altogether, all unwanted effects. To increase the chance of success, Münzenberg worked hard to optimise the SHIP.

Perhaps the most burning question was what proportion of reaction products would reach the detectors. Naturally, we did our sums and they even gave reasonable results, suggesting a yield of about 20%; for various reasons the rest would be lost. But calculations may be wrong, especially when based on uncertain assumptions with many inbuilt uncertainties.

1. The beam energy carries an uncertainty and it is also not entirely sharp, both factors being determined by the acceleration process and varying with the actual accelerator settings. The target is relatively thick and its internal structure may be inhomogeneous or may even change during the irradiation. The projectiles lose energy as they penetrate the target and we do not know where and at what depth the reaction would take place. The beam energy is high at the front face of the target but the reaction product has then to traverse the whole target thickness before it can reach our detector. At the rear face these considerations are reversed.

2. Another uncertainty lay in the fact that compound nuclei emit neutrons, protons or even α-particles with a spectrum of kinetic energies which has

to be approximated. Those emitted particles also influence the trajectory of the reaction product.

3. When the projectiles and the reaction products leave the target they do so carrying a distribution of charge states. Nobody knew the charge state of, say, element 114 when it leaves the target with a certain energy. That had to be approximated.

4. Energy, mass and charge distribution of the reaction products determine the trajectories through the SHIP. If the estimated values are badly wrong, a large fraction, possibly all, of reaction products may be lost.

Such uncertainties make it unreasonable to search for unknown superheavies over long periods: too much energy and money could be wasted to justify such experiments. Before starting a long search, a quick look seemed reasonable and might well give the results for which we hoped.

First, we wanted to prove the transmission calculation. To do so, we had to measure the number of reaction products directly behind the target and compare them with the number that we actually get behind the SHIP. But how could we measure behind the target? If this were possible we would not need the SHIP at all.

For the identification of fusion products near the target we tried gamma spectroscopy, chemical separation, detection of α-particles and fission fragments. We did indeed get data but they were not all as accurate as we would have liked, so our measurements of cross-section values were uncertain by a factor of two. However, this did not affect the measurement of excitation functions. Even now, we cannot yet accurately determine transmission values and finding a way of doing so remains urgent.

Suppression of background at the beginning of the SHIP operation was another essential task. First we had to locate the origin of the background. It was probably not due to the high energetic beam particles because these were well separated. Nor was the background coming from other reaction products. It seemed likely that the source was low energy projectiles which the SHIP could not separate because they had the same rigidities as the reaction products we wanted. The only solution was to avoid them.

What we did soon observe was that this background was not a function of beam intensity; sometimes even at high intensities the background was low. Where did it come from?

On one occasion we forgot to remove a grid used for positioning the beam. The irradiation had a tremendously high background. It was clear that part of the problem must have come from the accelerator itself. We started to optimise the beam trajectory and, from then on, things improved. Optimisation might take several hours, even a day or two, of beam time but it was a sacrifice we simply had to make.

The beam itself was also contributing to background so Münzenberg constructed a small velocity pre-filter, mounted 3 m in front of the target, to purify the beam by deflecting the low velocity projectiles. Sometimes it was worked well, sometimes not. The problem lay not with the filter but with variability of beam, so quite often the pre-filter was not used in experiments. Nowadays the beam quality is much improved and the pre-filter no longer needed.

The target itself, and the way the beam was focused on it, were also sources of background. Inhomogeneous target material used in manufacture, target frames too small and a beam halo which hit the target frame all contributed. But if the beam halo was avoided by focusing it more sharply, the target material melted, a problem we still have with lead and bismuth targets.

We naturally wanted to maximum possible beam intensity: that would give us the most product. However, the melting points of lead and bismuth are 327° and 271° Celcius. Why not use gold for example, which melts at 1,063°, tungsten (3,410°) or uranium (1,132°)? Because nature has arranged things so that only lead and bismuth have cross-sections promising for super-heavy element production. Here, as in so many other cases, experimentation is often a matter of compromise.

What could we do to be able to use lead and bismuth for the irradiation? The thickness of the target was governed by the reaction; the optimum density was 0.45 mg/cm^2, equivalent to a thickness of 0.4 micrometres (that is very thin, less than 1/2,000th of a millimetre or 1/150th of the thickness of a sheet of paper). The size of the beam spot on the target was governed by the ion optics of the SHIP; ideally it should not exceed 10 mm but the diameter of the target had to be considerably bigger to ensure the beam spot did not hit the frame.

Producing lead foils of 0.4 micrometre thickness and at least 15 mm diameter was not easy and it was Helmut Folger and his co-workers who found how to do so. He was a chemist; his target laboratory at GSI contained all sorts of rolling machines and high-vacuum coaters. The trick with the lead targets was to begin with a thin foil of carbon weighing only 40 microgram/cm^2 produced by sublimation (i.e. vaporisation and condensation) of carbon from a spectrographite rod onto a glass plate. These plates were previously treated with a solution of betaine-monohydrate, an organic compound which enabled the carbon to be stripped from the plates under water. The carbon foil floated to the surface of the water and was picked up on the target frame. After drying, the lead or bismuth was evaporated onto the carbon to the desired thickness. The target was finally coated with another thin carbon foil.

The carbon on both sides of the target hindered sputtering of the material during irradiation. It also helped to dissipate by radiation: with such very thin layers, heat conduction is negligible and cooling is possible only by radiation.

Over the years, Folger and his team made larger and larger targets. We were able to start developing target wheels with several target frames mounted on a wheel 365 mm in diameter. That gave a total length of the target material at the circumference of the targets of almost a metre. The wheel rotated through the beam at about 1,000 revolutions/minute thus distributing the beam intensity across the circumference of the target so that irradiated target spots had enough time to cool before they were irradiated again. Electron micrographs of new and worn out targets are shown in Figure 9.10.

One complication arose which demanded the special efforts of our electrical engineer Hans-Joachim Schött. Our beam particles were accelerated in macropulses of 4.5 milliseconds width every 20 milliseconds; that means at a frequency of 50 Hz, exactly the frequency of our AC mains supply. It was essential that the beam should never hit the spokes between the separate target panels on the wheel so the target wheel had to be rotated synchronously with the pulse structure ensuring that every time a pulse came down the SHIP it hit a target and not a spoke.

Schött arranged that, in principle, the wheel did run synchronously with the beam but, if you know how unstable this 50 Hz AC-frequency sometimes is, you will be not surprised to learn that the wheel did run out of phase from time to time, actually about once a night. When that happened, the beam was automatically shut down, the wheel stopped and the experiment waited for a human experimenter to intervene. The result was that night shifts became necessary; as a rule, people on shift had nothing else to do but wait for the wheel to stop or for a new element to announce its birth on the printout!

First success

Our first irradiation using a ^{40}Ar beam with ^{208}Pb targets mounted on the rotating wheel started on February 1st, 1980. After our earlier failures, we felt we needed to start again with something already understood. At the time, in Berkeley the synthesis of elements by hot fusion with actinide targets had reached $Z = 106$ while the Dubna team had made fermium by cold fusion using lead targets. At the UNILAC heavy ion accelerator, a very trivial reason excluded thorium and uranium targets: the low mass beam ions were too light. Beams lighter than argon were not allowed at the UNILAC because there were many accelerators worldwide capable of accelerating light

Figure 9.10 Scanning-electron micrograph of a new bismuth target (top) and from a bismuth target destroyed by the beam due to overheating (bottom). The melted bismuth formed droplets still sticking to the carbon backing foil; they were, however, too thick and thus useless for the heavy element synthesis. The scan with the electron beam across the destroyed target and detection of the emitted X-rays identified the droplets (bottom).

particles: it was felt that the UNILAC was so exceptional, it should be reserved for the really heavy species.

Lead and bismuth had problems because of their low melting points but elements below lead (thallium and mercury) are even worse. Gold and platinum were both possible targets but nothing was known about production of heavy elements using them; in addition, the number of neutrons is less favourable: ^{197}Au has eight and ^{198}Pt has six neutrons fewer than ^{208}Pb so lead was really the only serious choice.

In four days of beam time we had made ^{246}Fm, ^{245}Fm and ^{244}Fm, the first fermium isotopes produced in the SHIP, confirming earlier Dubna results using cold fusion. The α-spectra appeared very poor at the time but the lines were visible, the α-energy was where it ought to be and the daughter products (californium isotopes) were also there. The real progress was that we could individually correlate the decays of the decay chains using the new position-sensitive detector array.

A historically important experiment was performed from February 28th to March 4th, 1980, the irradiation of ^{208}Pb with ^{50}Ti resulting in the compound nucleus ^{258}Rf* (Z = 104). Here we wanted to make use of the advantage of the SHIP to measure short half-lives. The break of the systematic for fission half-lives at N = 152 and Z = 104 found by Dubna was still in question (see Figure 6.4). Our data came down in favour of Dubna: we measured a half-life of (8.1 $\pm$ 1.0) milliseconds for ^{256}Rf, close to the Dubna determination of about 5 milliseconds. The most recent value is (6.2 $\pm$ 0.2) milliseconds. In addition and for the first time, we synthesised the isotope ^{257}Rf which decayed by α-emission. This was the first observation that the nuclei of heavy elements can be produced at extremely low excitation energy, so low that only one neutron need be emitted to cool down the compound nucleus.

10

New elements and other stories: from bohrium to meitnerium

The interval between the syntheses of element 107 in February 1981 and element 112 in February 1996 is almost 15 years. The discovery of new elements is not equally distributed over this period: the first three were made in 1981–4, the last three from 1994–6. There is a reason behind this clustering of discoveries but we will follow events chronologically. For an easier orientation of where the new elements are located, how they decay and what their half-lives are, see Figure 10.1. This gives the most recent data up to the heaviest nuclei presently under investigation in a partial overview of the chart of nuclei.

Element 107 – bohrium

As early as 1976, the Dubna scientists reported searches for element 107, trying cold fusion and irradiated ^{209}Bi with ^{54}Cr ions and using rotating cylinder and fission track detectors for the identifications. Fission was indeed observed. One activity had a half-life of one to two milliseconds, the other five seconds. It was concluded that the short activity arose directly from the isotope ^{261}107, produced after emission of two neutrons from the compound nucleus ^{263}107 and the long one by α-branching of ^{261}107 into the daughter isotope ^{257}105 which was thought to decay by fission.

This was certainly a possible interpretation on the basis of the information available at the time although not a direct proof of the production of element 107. The interpretation was influenced by two facts:

1. the Dubna method was able to measure only fission; and
2. the break of the fission systematic at ^{256}104 led many experts to believe that for the heavier elements fission would be the strongest decay mode with shorter and shorter half-lives.

The GSI experts thought similarly. There was, however, the hope that for certain nuclei the fission process could be hindered so strongly that α-decay could dominate. The hindrance of fission was well known for lighter elements; it is observed for nuclei which carry an odd number of protons or

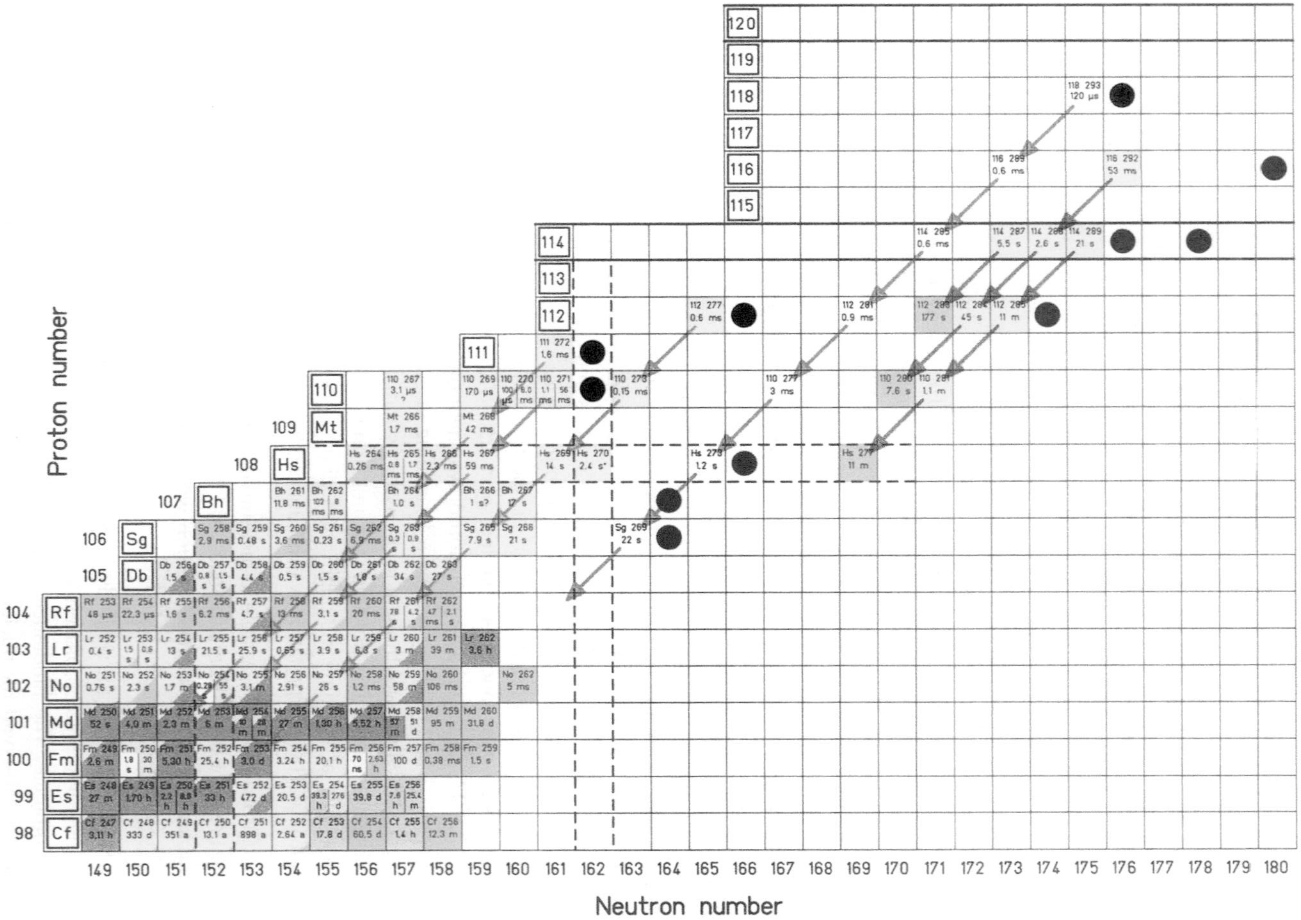

Neutron number

Proton number

neutrons because the odd particle requires additional energy to remain in its orbit when the nucleus deforms in a fission process. This additional energy (called *specialisation energy*) increases the fission barrier and thus prolongs the half-life. The surplus energy depends significantly on the spin value of the orbit occupied by the odd nucleon – the higher the spin, the higher the specialisation energy. Thus, the hindrance is not constant but varies roughly between 10 and 100,000; it averages about 1,000.

In the case of odd–odd nuclei, for which the number of protons and neutrons are both odd, the individual hindrance factors have to be multiplied so some odd–odd nuclei may have fission half-lives one million to one billion times longer than their even–even neighbours. Incidentally, that might be a useful pointer for finding superheavies in nature: search for odd–odd nuclei. While my expert colleagues might object that these nuclei are always

---

Figure 10.1 Upper end of the chart of nuclei showing the known nuclei and those which are presently (2002) under investigation. Yellow = α decay, red = electron capture or β$^+$ decay, blue = β$^-$ decay and green = spontaneous fission. The ratio of the areas in the case of multiple colours for an isotope corresponds to the branching ratio. For each known isotope the element name, mass number and half-life are given. The coloured dots mark compound nuclei of reactions which were used to study the heaviest elements. Reactions using ^{208}Pb target and beams of ^{64}Ni, ^{70}Zn for the study of elements 110 and 112 in Darmstadt are in blue. Element 111 was produced with a ^{64}Ni beam on a ^{209}Bi target. The isotopes of the new elements were identified by decay chains (nuclei connected with arrows) which ended in known, previously studied nuclei. The parent nucleus was created after evaporation of one neutron from the compound nucleus. Three long chains of seven subsequent α decays were reported from an experiment in Berkeley in irradiation of ^{208}Pb with ^{86}Kr. The chains were tentatively assigned to the superheavy nucleus ^{293}118, however, the result was retracted by the authors in July 2001. The red dots mark compound nuclei of elements 112, 114 and 116 studied in Dubna using a ^{48}Ca beam and targets of ^{238}U, ^{242}Pu, ^{244}Pu and ^{248}Cm. The observed decay chains are shown together with their assignment to superheavy nuclei created after evaporation of three and four neutrons, respectively. Finally, the dots in magenta mark compound nuclei used by nuclear chemists for the study of the chemical properties of elements seaborgium (Sg), bohrium (Bh) and hassium (Hs). The elements were synthesised in reactions with beams of ^{22}Ne and ^{26}Mg and targets of ^{248}Cm and ^{249}Bk. Relatively long-lived isotopes were produced after evaporation of four and five neutrons. The magic numbers for the protons at element 114 and 120 are emphasised. The bold dashed lines mark proton number 108 and neutron numbers 152 and 162. Nuclei with that number of protons or neutrons have increased stability; however, they are deformed like rugby balls or barrels contrary to the spherical superheavy nuclei.

β emitters, it remains true that some of the odd–odd nuclei close to the line of β stability (marked by the black squares in Figure 1.5) have β half-lives of more than a million years.

Back to element 107, which is also an odd element, with the isotope $^{262}107$ an odd–odd nucleus with 155 neutrons. We therefore hoped that the fission half-life would be so hindered that we could be able to identify $^{262}107$ by α-decay and make use of our newly developed position-time correlation method. But correlate to what? The daughter product $^{258}105$ was unknown at the time. Should we start with 107? We decided not to.

Although some people thought it would be a waste of beam time, we wanted first to identify the daughter using the reaction $^{50}Ti + ^{209}Bi \rightarrow ^{258}105 + 1n$. It was a five-day experiment in February 1981. We detected 129 nuclei and measured the half-life as 4.4 seconds, found the α-decay energies and also observed ^{254}Lr, the granddaughter nucleus of $^{262}107$, which was unknown at the time. The α-decay chains ended (for our detection method) at the known isotopes ^{246}Es or ^{242}Bk due to the strong β-decay branches of these nuclei or, in some cases, at ^{250}Fm and ^{246}Cf because of the long half-lives of these nuclei. Detector signals from background particles prevent half-lives longer than several minutes being safely assigned to the decay chain. The nuclei ^{250}Fm and ^{246}Cf were produced due to β-decay branches of ^{254}Lr and ^{250}Md. The decay pattern as it is known today can be easily constructed from Figure 10.1

A very important observation was that of fission but that could be excluded for the decay of $^{258}105$ itself. The measured fission events were due to a 33% electron capture decay branch of $^{258}105$ into $^{258}104$; this isotope has an almost 100% fissioning nucleus with the short half-life of 13 milliseconds. Therefore the fission activity was measured essentially with the same half-life of 4.4 seconds as the α-decay of $^{258}105$. Unfortunately, β-decay and electron capture are generally not picked up by our detector.

At a slightly higher beam energy we were also looking for the neighbouring lighter isotope $^{257}105$. We expected that at higher beam energy another neutron would be emitted from the hotter compound nucleus, so producing $^{257}105$. This was the one that the Dubna people claimed to have seen as a daughter of $^{261}107$. According to their data, $^{257}105$ should have a half-life of five seconds and a fission branching of 20%. We did indeed measure a few fission events but our value for the half-life was 1.4 seconds, so the five second activity could not have been due to $^{257}105$.

The five-days titanium beam run we did before the main experiment was a good investment, yielding four new isotopes: $^{258}105$ and its daughter $^{254}103$, $^{257}105$ and its daughter $^{253}103$. We correctly assigned their half-lives and knew the decay branchings. Well prepared, we started the main experiment on February 22, 1981.

The beam time allocation was only four days, even shorter than for the preparatory experiment – and that for an attempt to produce a new element. It seems to me now that we were not brave enough. But we got a two-day extension because there was a delay at the beginning of the run. The beam was ^{54}Cr, the target ^{209}Bi, the compound nucleus ^{263}107 and the beam energy was so adjusted that one neutron would most likely be emitted. That should give ^{262}107 which would hopefully not fission but α-decay into the daughter nucleus ^{258}105, already identified.

The first sign of 107 was the rattling of a typewriter during the night shift on February 25th, 1981. From today's viewpoint this sounds like the Stone Age but it was 21 years ago, the PC age had not yet started, computers were extremely expensive (only institutes could afford them) and their capacity was small. Our PDP machine from Digital had a magnetic core memory of only 40 kilobytes so all the programming had to be very condensed.

The main purpose of our experimental computer was to collect the numbers from the analogue-to-digital converters and scalers, and store them event-by-event on magnetic tape. These tapes were later analysed off-line using the larger central computer.

A small program on the on-line computer told us that the experiment and data collection were running well. However, a complete correlation analysis was not possible on-line. To do that one had to store a certain number of events using a large proportion of the memory. For each new signal from the detector, the preceding ones had to be recalled to determine whether they were from the same detector strip at the same position as the new event. If so, there was a good chance that the two events were parent and daughter decay. The analysis could be extended by distinguishing between signals originating from ions that passed SHIP and were implanted into the detector and those from the radioactive decay of the previously implanted nuclei. To do that we used two additional detectors mounted in front of the silicon stop detector. These were made of thin carbon foils through which the ions could easily pass but which delivered a signal whenever one went by so we called them *transmission detectors.* Their signals were registered in additional branches of the data acquisition system together with those from the silicon stop detector. The transmission detectors do not fire for radioactive decays and we get signals only from the stop detector. A whole collection of signals characterising *an event* appears almost simultaneously. Their appearance and height indicate implantation or radioactive decay. However, all that information was for off-line analysis.

On-line, I programmed a very simple but nevertheless effective method of detecting correlations. It worked only for those events which happened during the beam-off period. This is the period of 15.5 milliseconds between the beam pulses which are formed every 20 milliseconds for a duration of 4.5 milliseconds. Whenever an α-signal had an energy higher than, say, 8 MeV, the three most important informational items about the event were printed on a teletype. The same thing was done for a fission event. The information collected included the height of the detector signal (related to the decay energy), the position in the detector and the time in units of 160 milliseconds for reasons which will be explained when we look at the mechanics of the instrumentation. Figure 10.2 shows such a printout, actually the first time we saw element 107. Because the printing of the teletype was very noisy, each output sounded like an alarm clock, alerting the curious researchers to the possible birth of a new element.

The page begins with the opening of a new file on the magnetic tape connected to the PDP machine: tape 4, file 3 of SHIP run 71. The irradiation for this file started with the 'GO' command at 01:03:08 (a.m.) on February 25th, 1981. There follow a number of display commands which were typed into the machine; they all started with the letter 'D'. At 1:04 a.m., one and a half minutes after 'GO', one of the spectra accumulated for the on-line control was cleared, which means it was reset to start. That had no influence on data collection but was necessary from time to time for controlling the progress and stability of the set up.

Later on, two lines were printed which caught the attention of the two people on shift, Willi Reisdorf and Karl-Heinz Schmidt. They soon realised that they signified something special. Using the known energy calibration, they read off the energies from the channel numbers 2466 and 2075 as 10.40 and 7.73 MeV, respectively. The next number determines the position in the detector, 338 and 340: the hundreds stand for the detector number (no. 3 in these instances) while the tens and the units give the vertical position. The overall detector length of 27 mm was 63. The two events were 0.9 mm apart.

The third number gives the time. The strange time unit of 160 milliseconds was based on the fact that we used a rotating wheel which carried 16 targets (the new version has only eight, however, larger targets). The 16 targets were subdivided into two groups of eight targets each. If one group were destroyed, we could shift the phase of the rotation 22.5° to the second group. It sounds complicated but the length of the targets were such that there was room for two groups. When one group of targets was eventually destroyed by the beam current, we could shift to the other and thus not have to open the vacuum chamber to replace them. That saved about three hours every other time we had to change targets.

The natural time unit in our system was determined by the UNILAC

```
25FEB81  01:02:59 FILE 3 OPENED ON PDP-TAPE R71AT4
*GD
GD 25FEB81 01:03:08
LIST MODE DATA WILL BE DUMPED ON PDP TAPE
*DLI
*D 2
 D/LIVE
*ACL 2 1
CLEAR ANALYZER 2 OF EVENT 1 (Y/N)?:Y
CLEARED AT 01:04:44
 D/LIVE
 D/LIVE
*DLI 2 1

2120., 216., 1212.,
*D 28
*DLI 26 1
*D 26
*DSC 100
*DLI 2 1
*DLI 22 1
*D 35
*DLI 35 1
*SD 1

!GOCO: UNKNOWN COMMAND!
*D 1
*D 28
 DWSUM
 :5056-5559=1
*D 40
DSPL:SPECIFIED SPECTRUM RANGE IS EMPTY
*D 5
*D 26
*DSC 100
*D 35

2466., 338., 12820.,
*D 40
DSPL:SPECIFIED SPECTRUM RANGE IS EMPTY
*D 37                                        ΔT = 219 sec
DSPL:SPECIFIED SPECTRUM RANGE IS EMPTY
*D 5
*D 30
*D 28
 DEXP
*D 31

2075., 340., 14194.,
*D 28
 DEXP +
*D 5
*D 28
```

Figure 10.2 Computer printout from the night February 24–25, 1981, with the first identification of element 107.

pulsing at 50 Hz, one pulse every 20 milliseconds. For the eight targets we scaled this unit down by a factor of eight, giving 160 milliseconds, the strange time unit we noted above. Within 160 milliseconds all eight targets were irradiated, then the target cycle started again. However, the finest subdivision for the time measurement was much less at 1 microsecond. We achieved this by increasing an electronic scaler by one every microsecond and resetting the scaler at the end of each 50 Hz cycle. We counted the number of cycles in a similar fashion. The contents of both scalers were read out together with each detector signal and written to tape. It resulted in fixing the time for each event with an accuracy of 1 microsecond, an important property of our set-up for measuring short lifetimes.

With that arrangement we could easily determine the time difference between the two events as 219 sec. Willi and Karl-Heinz had worked that out during the night. They looked up the tables and found that the second event was a good candidate for the decay of ^{250}Md. The first α-energy was relatively high and unknown; could it have been the decay of ^{262}107?

They decided to remove the recording tape and analyse the event 'off-line'. Naturally they replaced it with a new one and continued collecting data. However, changing the tapes took about 15 minutes. Later on they regretted not waiting for the decay of ^{246}Es which has a half-life of 7.7 minutes and 9.9% α-branching. Observing the ^{250}Md α-particle, by the way, was a lucky break as its α-branching is only 7%. Perhaps one should not expect too much from one event.

The overnight off-line analysis suggested a preliminary chain for the decay of ^{262}107 (Figure 10.3). Then it was champagne time again and they let off fireworks which they had kept from a previous New Year's Eve celebration. Next morning the rest of us were able to take a look; we all agreed with the explanation the champagne drinkers had given during the night.

The experiment ended on February 28th and off-line analyses started immediately. We detected a total of six decay chains; all were assigned to the decay of ^{262}107. The first and second had registered at 10:48 a.m. and 7:24 p.m. on February 24th but this had either not been printed on the teletype or had been overlooked. Five chains started with a short lifetime of 6.9 milliseconds, another of 165 milliseconds. We concluded that two different energy levels must have been produced in ^{262}107, a finding confirmed later.

Measuring half-lives does not necessarily mean waiting until half of the activity has decayed because, with enough activity, the rate against time can be plotted and the half-life simply calculated. That, of course, is difficult if only one atom is produced! However, even with a single atom we can determine the lifetime of a nucleus. We know when it was produced and when it decayed: the interval is the lifetime, and from that we can calculate the half-life by multiplying by 0.69 (this is the natural logarithm of 2; see also Box 1.2 in Chapter 1). For a single atom this is, of course, no more than a mathematical number but nevertheless it is correct within relatively large error bars for any larger number of atoms of the same type. The time values shown in Figure 10.3 are the measured intervals between the detector signals and

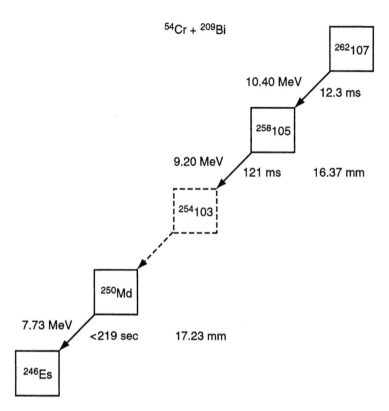

25.2.1981 1:30 h

^{54}Cr + ^{209}Bi

^{262}107

10.40 MeV

12.3 ms

^{258}105

9.20 MeV

121 ms   16.37 mm

^{254}103

^{250}Md

7.73 MeV

<219 sec   17.23 mm

^{246}Es

Figure 10.3 Element 107, bohrium, the first element made by a cold fusion reaction was synthesised by the reaction ^{54}Cr + ^{209}Bi → ^{262}Bh + 1n. This diagram, dating from the early morning of February 25, 1981, shows the tentative interpretation of the event chain observed at 1:30 a.m. It was detected with detector number 3 from the left at a position 16.37 mm from the bottom. The new element appeared as an α-particle of 10.40 MeV energy correlated with an evaporation residue impinging on the detector. It was also correlated with a 9.20 MeV α-particle occurring 121 milliseconds later at the same position, attributed to the known ^{258}105. The grand-daughter ^{254}103 was not seen, because the α particle escaped from the detector in a backward direction, whereas the last chain member, ^{250}Md, followed after 219 seconds. The '<' sign takes into account that for the lifetime of ^{250}Md the literature value of 19 seconds for the lifetime of ^{254}103 has to be subtracted.

thus the lifetimes. The chart of nuclei, however, shows the half-lives commonly used.

In a fusion reaction the nucleus is actually produced in the target. The first we know of it is when it is implanted in the detector. By then it has already lived two microseconds so that has to be added for very short lifetimes. When lifetimes are in the millisecond range, this effect is negligible. The lifetime of the daughter nucleus starts when the mother decays and so on along a decay chain. We get the times from the signals for α-decay or fission.

One more interesting aspect arises when a nucleus has two or more ways of decaying, say by α-decay and fission or by α-decay and electron capture. Then we speak about *branching ratios*: how much goes by one route and how much by another? In the isotope ^{258}105, 33% undergoes electron capture and 67% α-decay. The lifetime is thus determined by both decay modes and the total or mean lifetime is 6.3 seconds (to get the half-life, multiply by 0.69); that is the experimental value. But the individual lifetimes (*partial lifetimes*) are longer: they can be derived by dividing the total lifetime by the branching ratio. Thus, in the case of ^{258}105 the partial α-lifetime is 6.3 seconds/0.67 = 9.4 seconds while that for electron capture is 6.3 seconds/0.33 = 19 seconds. If the partial lifetime is very long, the resulting branching is very small; the branchings themselves are worked out from the relative intensity of the decay modes, i.e. number of α-particles compared with the sum of electron capture events plus α-particles. Here again we have this interesting phenomenon that lifetimes determine the intensities. We already have alluded to it in Chapter 8 when we discussed the probability of neutron emission and fission of a hot compound nucleus, exactly the same relationship.

Within four weeks, data analysis on ^{262}107 was complete and a short paper submitted for publication which appeared in the *Zeitschrift für Physik* as 'Identification of Element 107 by α Correlation Chains'; Münzenberg, the first author of the paper, also presented the findings in June 1981 at an international conference on *Nuclei far from Stability* at Helsingør in Denmark. People were present both from Dubna and Berkeley but none of them spoke on heavy element research. We were out in front!

The experiment ^{54}Cr + ^{209}Bi was repeated over six days in March 1985 and again over 11 days in August 1987. We were aiming to synthesise the neighbouring isotope ^{261}107 because it was still unclear whether the Dubna people were right in assigning their 1–2 milliseconds fission activity to the decay of this particular isotope. If they were, they would have priority for its discovery but if not it might fall to us. The second objective was to improve the quality of the ^{272}107 decay data. The statistical errors for lifetime measurements, and especially the upper limits for fission, for the small number of events detected were still too large for comfort. And we also wanted to confirm the one event which had that long 165 milliseconds lifetime.

The isotope $^{261}107$ was clearly identified by a total of nine measured decay chains. We determined the half-life as 9.0 milliseconds, significantly longer than the Dubna value (the latest value is 11.8 milliseconds). Two fission events were attributed to a fission branch of the daughter $^{257}105$, because they were preceded by an α-decay. We concluded that the Dubna data must be erroneous and the question of priority for discovering of element 107 was solved. The researchers in Dubna agreed with our findings. In order to deepen our good relationships and to acknowledge the preceding work at Dubna, we later invited Oganessian to become the godfather for this element – which he accepted with pleasure.

There now existed 29 decay chains for the isotope $^{262}107$, 14 of them were giving a half-life of 8.2 milliseconds for the short lived level and 15 indicating one of 106 milliseconds for the longer lived level, probably the ground-state of $^{262}107$. In neither case could fission be attributed directly to the decay of $^{262}107$. The production cross-section (see Chapter 8) was 99 picobarn for the ground-state. With present day (2002) technology, this implies a production rate of about 40 atoms per day.

Further decays of $^{262}107$ were found in August 1982 and February 1997, when $^{262}107$ was produced as a daughter of $^{266}109$. Finally, one new isotope ($^{264}107$) was identified in December 1994 as a granddaughter in the decay chain of element 111. It was measured again in a confirmation experiment for element 111 in October 2000. This more neutron rich isotope has a half-life of 1.0 seconds and is an α-emitter. Note the increase in half-life from 106 milliseconds to 1.0 seconds when the neutron number changes from 155 to 157. Longer half-lives correlate with higher stability. This means that with heavier isotopes we are approaching a region of higher nuclear stability. But this was a concept we did not work out on 1981 when we had produced just one isotope, $^{262}107$. Using the reaction $^{22}Ne + {}^{249}Bk \rightarrow {}^{271}107^*$, our colleagues in nuclear chemistry at Zurich and at Berkeley have very recently (1999) added two more neutron rich isotopes with even longer half-lives, $^{266}107$ (1 sec) and $^{267}107$ (17 sec). The official name for element 107 is bohrium (Bh) after the Danish physicist Niels Bohr (Figure 10.4)

Element 109 – meitnerium

Back in April 1981, after the element 107 run, I was extremely busy at the SHIP, not then with element 109 but preparing one search for proton radioactivity and another for element 114 – yes, that is correct: 114. The first experiment was a success, the second was not. These activities filled my time

Figure 10.4 Niels Hendrik David Bohr, October 7, 1885–November 18, 1962: in his honour, element 107 was named bohrium (Bh). Bohr was born in Copenhagen, became a lecturer in Manchester in 1914, professor at the University in Copenhagen in 1916 and Director of the Institute for Theoretical Physics in 1920. He received the Nobel Prize in 1922.

until the 109 experiment in August 1982; they were important and we learned a lot from them.

We were successful in finding for proton radioactivity: ^{151}Lu was identified as the first nucleus emitting protons from the ground-state, a nucleus located far from the β-stability line. That merited another paper for the Helsingør conference. In the first part of the experiment we were unable to measure the half-life of the proton emitter because we simply did not have enough beam time and we had to go through all the bureaucracy to get more, so did not do the next experiment until late July. We found a half-life of 85 milliseconds, got an improved yield for the production of ^{151}Lu and could demonstrate convincingly that the line in the spectrum really did come from protons: combining thin and thick silicon detectors to form a telescope enables clear distinction between lightly-charged particles like electrons, protons or α-particles.

This type of experiment is much more work-intensive than a heavy element run. There were many targets to irradiate, the detectors were more complicated and the amount of data to be analysed much greater. Nevertheless, working in various areas of the nuclear chart was generally worth it; we learned how to operate the SHIP, the detector systems and the analyses more effectively and tried to combine all our results into an understanding of the underlying physics.

Each experiment has to be analysed and eventually published. If that is not done immediately, the data are usually lost; one of the problems is that not everything is on paper or in computer records – particularly some of the thinking that might have gone into the experimental design and operation – so the memories of the experimenters play an important part in shaping the write-up. I regret that I have one case in preparation ongoing for 15 years; I hoped to get it done in the year 2000. However, after the July 1981 proton run I did make time to analyse the data and wrote a paper.

To be consistent and work stepwise in a logical manner is always easy – but sometimes the gambling instinct takes over. This was the case in an experiment in May 1982: the reaction was ^{48}Ca + ^{244}Pu → ^{292}114*. With the new efficient detector, and after the successful 107 experiment, we thought there might be a chance for finding 114. We were also fortunate in persuading Al Ghiorso to participate and to supply the target materials, two of ^{244}Pu, one 0.18 and the other 0.27 mg/cm² thick, electroplated on 0.25 mg/cm² titanium backing, and extremely poisonous.

In this experiment the two reaction partners were very special. The projectile ^{48}Ca is a very rare isotope: in natural calcium its abundance is only 0.19% so to get a strong enough beam we had to use enriched material. That is extremely expensive, much more so than gold: at the time one gram cost about $300,000! Nowadays it is 'cheaper' at 'only' $110,000 because the enrichment is no longer carried out in magnetic separators but in centrifuges. Not surprisingly, our ion source people went to great lengths to keep the consumption of material as low as possible. They managed to get it down to 2 mg per hour, 200 mg during the four days of beam time. Cheap, really, all things considered: only $60,000!

Special care was necessary with the target. 'What would happen if . . .?' was the inevitable question of our safety people. So we tried to anticipate every eventuality but I am sure that, had something happened, we would not have been prepared. Luckily, nothing did go wrong but little went right either. We will come back later to the reaction $^{48}Ca + ^{244}Pu \rightarrow ^{292}114^*$; it is the one which again created great excitement when used recently at Dubna in searching for (and detecting) element 114.

This element 109 experiment was the most 'beautiful' of the three (107, 109 and 108) of the first period. I call it beautiful because everything fitted together (almost) perfectly: the preparation, preparatory experiment, the main experiment, the result and the fast publication.

You may ask why we went for element 109 before 108? There were two reasons:

1. isotopes of element 108 were still expected to fission, possibly so fast that we would not be able to detect them. You may remember that fission should be hindered for odd elements and fission half-lives should be especially long for odd–odd isotopes;
2. there exists a wonderful chain of possible projectile nuclei, ^{50}Ti, ^{54}Cr, and ^{58}Fe which differ by α-particles. When using ^{209}Bi as a target, the compound nuclei ($^{259}105$, $^{263}107$ and $^{267}109$) and the residues after neutron evaporation differ also by α-particles. We can therefore neatly climb the element ladder in steps of α-particles, using those projectiles, and slide down it by α-decay. Such decay of the new isotope will populate the known chain from the previous experiment. By the way, the ladder continues with ^{62}Ni and is valid also for even elements using ^{208}Pb target (Figure 10.5). This property was important later in searching for element 110.

However, like so many good things, these isotopes are scarce. The least abundant is ^{58}Fe forming only 0.28% of natural iron. We needed enriched material in order to get enough high currents and, as in the case of ^{48}Ca, special care and preparation was mandatory.

We ran the preparatory experiment, $^{50}Ti + ^{209}Bi \rightarrow ^{258}105 + 1n$, over three days in August 1982. This irradiation was expected to provide more accurate data for the decay of the granddaughter $^{258}105$ and also reveal any

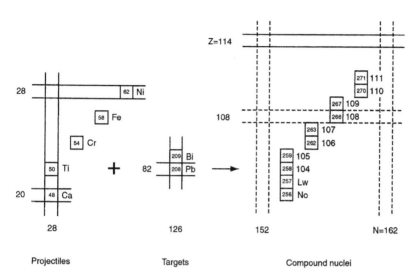

Figure 10.5 Possible reactions for the synthesis of heavy elements from nobelium to 111 using targets of lead and bismuth. Full lines indicate closed shells, dashed lines neutron numbers of deformed nuclei exhibiting maximum local stability. The crossings at $Z = 114$ indicate that, presently, it is not known if these nuclei are spherical or deformed.

flaws in the detection electronics. The experiment did run extremely well, so well that some people feared it might be a bad omen for 109! We spotted 228 α-decays of $^{258}105$ and confirmed the electron capture branch to $^{258}104$, the fissioning isotope (see Figure 10.1), by detection of 135 fission events.

Then the problems started. One of the accelerator's huge pumps failed; it was essential for pumping the nitrogen gas of the gas-jet converter after the first accelerator section. Could the gas converter be replaced by a solid carbon foil stripper? In that case we would not need a pump at all. But that was impossible; the currents were too high. Hectic activity spread among the accelerator people. We decided that that must be a good omen!

The choice of the beam energy for the main experiment was also giving us headaches, so the enforced pause allowed us another chance to discuss it. The arguments flowed back and forth, too nicely balanced to be conclusive. In the end we came to a decision, quite how I am not sure. The pump was repaired, the run commenced but after five days of irradiation there was still no event. For four days we lowered the beam energy so that the temperature of the compound nucleus decreased by 14%, but still there was nothing. As a last desperate step we raised the beam energy again, higher than the original energy; the temperature of the compound nucleus was now 12% higher than when we started and our scheduled beam time was already over. However, we had lost three days at the beginning and so asked for more time. The bureaucracy was sympathetic and allocated four additional days. We decided to stay with the last higher beam energy using the simple argument that more must be better than less.

And so to Sunday, August 29, 1982: 15 days of beam time behind us. Although my shift did not start until midnight, for some reason I was present that afternoon. At 16:03:19, so the record states, I closed file no. 7 of tape 15. It was a measurement taken during the change of the ion source when there was no beam. Nevertheless, we collected the data to have a control for the background and to detect any long-lived daughter decays. At 16:04:15 the new file (no. 8) was opened with the comment *'WEITER FE AUF BI 5.15 MeV/U NACH Q-WECHSEL'* i.e. 'continuing Fe [that was the beam] on Bi [the target material] 5.15 MeV/nucleon [beam energy] after change of the ion source'). The GO command was given at 16:05:57.

Shortly after, the teletype rattled: two lines were printed within a few seconds, an indication that something special had happened. The discriminator levels for the printout were set fairly high so the typewriter rattled relatively infrequently. Sometimes it got on our nerves, especially after 15 days of beam time and still no event. But this time it was interesting. A fission event was signalled, preceded by an α of high energy so I drew a big question mark close to the lines on the printout. This time we did not remove the tape. Although no more decays were expected after the fission, it would be good to have a more complete range written to tape before and after this special event. Only a short period of 2 minutes and 39 seconds was missing when the file was started.

I no longer remember exactly how things developed. I guess that some time that evening, or during the night, Willi Reisdorf analysed the data tape off-line. He also found the missing link, a 1.14 MeV escape α which was not printed because of its low energy. Then Karl-Heinz plotted the full decay chain; it looked very convincing. It was written to tape at 4:10 p.m., four minutes after restart of the experiment. What would have happened if, during the opening of the new file, somebody, with high probability our director (usually directors have time only on Sunday afternoons), had asked a question which took four minutes to answer?

That fission event was the only one observed during the whole experiment with the ^{58}Fe beam during the pulse pauses. The energy agreed well with the energy of fission events measured with the ^{50}Ti beam due to fissioning of $^{258}104$ and the lifetime of 22.3 milliseconds agreed well with the mean value of the five events of $^{262}107$ measured a year and a half earlier. Unfortunately the α-particle missed the detector and we got only a fraction of the full energy. The signal height, however, corresponded to that of an escaped α-particle, and we had the time of the decay. The energy and time of the preceding α-particle showed it must have been emitted from $^{266}109$.

Within four weeks of ending the experiment, all the data were analysed off-line. No more events were found so the title of our short note to *Zeitschrift für Physik* was 'Observation of one correlated α-decay in the reaction ^{58}Fe on ^{209}Bi $\rightarrow$ $^{267}109$'. That fairly described what we had.

One year later we had written the full paper (14 pages long) discussing all the pros and cons for the assignment. At the end of the paper, in the section on our conclusions, we wrote: '*A fortiori*, our hopes of making still heavier elements, such as element 111 by using ^{64}Ni projectiles, are small since we have obviously reached the limit of production cross-section ($\cong 10$ picobarn) that can be observed using present day techniques . . .'. Quite true, as it turned out.

The experiment was repeated using an experimental set-up identical to the first experiment in January/February 1988 when two more chains were measured. The second of these, however, started with the decay of $^{262}107$. There were reasonable arguments for the loss of the signal from the $^{266}109$ decay, including thresholds in the electronics, but because $^{266}109$ was not observed, we cannot count that event as detection of element 109. Also in the first case we were lucky only partially. Here, the $^{266}109$ α escaped, however, the chain could be followed down to ^{250}Fm. With the second experiment we satisfied the objections of some critics that the first experiment should be repeated and that one chain is not enough evidence for the discovery of a new element.

In honour of Lise Meitner (Figure 10.6) we proposed the name *meitnerium* (Mt) for element 109. An excerpt from our letter to IUPAC and IUPAP said: 'Lise Meitner is one of the outstanding women in nuclear science to whom we feel especially obliged. She was one of Germany's most eminent physicists and was forced to emigrate in 1938, shortly before the discovery of fission by Hahn and Straßmann, whose work she initiated.'

Twelve more decay chains of ^{266}Mt were measured in a recent experiment (February 1997) aiming at an excitation function for the production of ^{266}Mt. The new data revealed a complicated α-decay pattern and an improved half-life value of 1.7 milliseconds. In December 1994, a new isotope, ^{268}Mt, was measured as a daughter nuclide after α-decay of $^{272}111$. This isotope was observed again in the October 2000 experiment for the confirmation of element 111. From a total of 6 measured decays a half-life of 42 ms was determined.

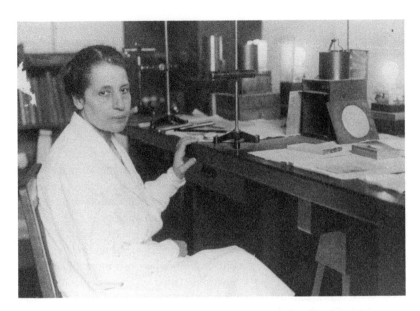

Figure 10.6 Lise Meitner. In her honour, element 109 was named meitnerium (Mt).

Important visitors and wanderings from the path of virtue

The discovery of a new element is not only a great event for the scientists and the scientific community. It is also a great social event for the whole institute and the broader public. What should one do when a new element is identified? Who should be informed and how? Should we wait and, if so, for how long? People naturally want to discuss exciting new results with colleagues and to spread the news. But what is the correct procedure particularly as we are government employees and there are rules. One question above all others has priority: is the 'result' correct? Nobody can help us answer; it is our responsibility and our right, that of the scientists who performed the experiment.

At the present time, the discovery of a new element has no commercial implications (although who knows what might develop in the future?); there are no patent rights to be considered. Luckily, we can concentrate on the physical results and the science of our work.

It is always the obligation of scientists to report their work and doing so invariably clears the mind. All detailed experiment details have to be assembled and correlated, the results worked out, possible flaws checked and misinterpretations considered. Then the findings and conclusions are discussed in the context of current knowledge, their value assessed for scientific understanding and a view taken of how the experimentation might be improved and how the field might go forward. All this forces us to think through all the implications of our work, gives us time for ideas to mature and to convince ourselves that we have got it right; publishing papers is also a major route for communication with our scientific colleagues.

When a paper has been prepared and submitted, it may be time to inform the public. In a scientific paper one can argue possibilities and multiple interpretations but that is quite impossible in a press release; there everything must be in black and white or the exercise gets out of control. We learned that for the first time with element 107, and for the second time when recording a single event for element 109.

I remember some cases which did run wild. In June 1976, element 126 was 'discovered' in the US, which later proved to be false. Then there was 'cold fusion'. Every physicist will remember where he was in April 1989 when this story hit the headlines in the media. I happened to be in Poland at the Zakopane School of Physics. Cold fusion appeared to be a classic case of what the Nobel chemist Irving Langmuir called 'pathological science' in which the results are always near the limit of detectability and the proponents always have an *ad hoc* answer as to why that should be so. So it was in 1982 that colleagues at Berkeley and Dubna may have had doubts about *our* discoveries; some of them came to Darmstadt in order to be convinced.

Our first visiting VIP was Glenn Seaborg, the 1951 Nobel Laureate in Chemistry; he came in September 1982, a great man, tall and serious, and seventy years old. Figure 10.7 shows him together with Gottfried Münzenberg and me as we explained our correlation method.

Next, in December 1982, came Georgi Nikolaevich Flerov. He was a Lenin Prize Winner and Academician of the Russian Academy of Sciences, also a great man but younger, not so tall and not nearly so serious (Figure 10.8). I had already met him at a conference in 1976 and I think he also remembered me from the Helsingør conference a year earlier when Gottfried had presented our work on element 107 and I talked about the proton emitter ^{151}Lu. Flerov taught us how to synthesise superheavy elements and suggested how we should continue our work. Eventually we received invitations to carry out some experiments at Dubna.

From then on, with close personal contacts, our relations with both laboratories became even more friendly than they had been previously. The growing friendship resulted in a common experiment with the Berkeley group: we decided to work on the reaction ^{48}Ca + ^{248}Cm → ^{296}116*, our deviation from the path of virtue. As we now know, it was much too early.

Ca plus Cm was – and still is – the dream of superheavy element researchers. The fusion of both nuclei would provide element 116 and an

Figure 10.7 From the left: Münzenberg, the author and Seaborg at GSI discussing the results of the 109 experiment, September 22, 1982.

isotope with neutron number 180, close to the magic number of 184. However, as we have already noted, ^{48}Ca is rare, while ^{248}Cm is a highly radioactive synthetic element with a half-life is 3.4×10^5 years; it is available in the US in 10–100 mg quantities.

The reaction had already been investigated in the mid-1970s both at Berkeley and at Dubna but with no results. At the beginning of the 1980s the idea came up again for three reasons:

1. the SHIP's performance in separating fusion products had been excellent. In the meantime, Ghiorso had built his gas-filled separator SASSY (Small Angle Separator SYstem) at Berkeley;
2. the in-flight separation and implantation of the reaction products into position-sensitive detectors had extended the accessible lifetime range by at least three orders of magnitude down to microseconds. On the other hand, the new detectors allowed the measurement of long decay chains with half-lives up to almost an hour, as had been demonstrated in the element 107 experiment. Chemical separation and detection methods

Figure 10.8 From the left: Münzenberg, the author and Flerov joking about fusion; GSI, December 1982.

had been developed which covered a half-life range from seconds up to several years;

3. the reaction process had been reinvestigated theoretically and the new results indicated that the beam energy could have been too high in the early experiment. It would be a good idea to repeat the experiment at lower beam energy. Again we can see how crucial is the choice of beam energy in heavy ion fusion reactions (and not only there). The new insight came from a recalculation of the extrapush which will be explained later on.

Our first plans for a common experiment with the people in Berkeley were made in early summer 1982. It was not only the recoil separator physicists who participated; there were the groups of nuclear chemists from Berkeley, Livermore, Los Alamos, Bern, Mainz, Darmstadt and Göttingen. Chemically, element 116 belongs to group VIA and so might be expected to behave like sulphur, selenium, tellurium and polonium. Would this progression be disturbed for element 116 by the tight binding of the electrons to this high Z nucleus? By how much? Answering this question was the basic scientific aim of the chemists and of vital importance for detection by chemical separation. The question is still unresolved.

In October 1982, the first part of the experimental programme started in Berkeley. Together with Gottfried, I embarked on my first flight to San Francisco and found a room in a wonderful house in the Berkeley hills; from the deck there was a panoramic view of San Francisco Bay, the Bay and Golden Gate Bridges and far out over the Pacific Ocean. The sunsets were unforgettable and I am still grateful to my hosts for letting me stay with them for three weeks. But in Germany I had heard that it never rains in California so I left my umbrella at home. That certainly was a mistake.

The Lufthansa flight arrived in San Francisco at noon, about 10 p.m. for people who had woken up that morning in Europe, so almost a full day lay ahead of us (or night, depending on how you looked at it). We went to the Berkeley lab. to get a first impression. It was very different from what we knew at home. Our institute was new and there was still lots of empty space in the experimental areas but in Berkeley we could touch history. We walked the length of the SuperHILAC and, through a glass window, watched the beam, ionised and shining in the stripper region. The vertical wheel which, in 1974, was an essential feature in the discovery of element 106, was stored close to Ghiorso's cave. In accelerator physics we call the experimental areas 'caves' because they are surrounded by concrete walls for radiation shielding (but believe me, physicists do not behave like cavemen – usually).

We saw Ghiorso's records – twenty or more volumes of them side by side on a shelf, each one about an inch and a half thick and bound in dark green linen with leather spines. All the experimental details were recorded in small clear handwriting, the life of a great scientist.

The chemists had already started their irradiation when we arrived. It was a good time for us physicists to learn something about the advantages but also the problems of 'radio'-chemistry. The difference from normal chemistry is that the stuff the radiochemists deal with is, of course, radioactive, sometimes has a short half-life – extremely short handling by normal chemical techniques – and, in our case, produced in tiny quantities: we needed 'one atom at a time' chemistry. This does not mean that radiation safety is not a problem. Huge amounts of radioactive isotopes are produced, much of it of lighter elements which have to be separated. This is the first step in the procedure; the next is to isolate and identify the superheavy element which hopefully has been produced – one atom at a time.

I want to give two examples which show that radiochemistry does not necessarily have anything to do with solutions mixed in test tubes. Common to the on-line separations was a helium-jet transport system. It included the target, a stopping chamber for the reaction products and a capillary for transporting the activity to the chemical separation apparatus. Such a transport system, shown in Figure 5.1, had already been used for the identification of element 106 (seaborgium). The transport time was two seconds. Expecting element 116 and its daughter elements after α-decay down to element 112 to be volatile at 1,000 °C (just like lead), the chemistry set-up shown in Figure 10.9 was used: a continuous high temperature gas chromatograph. As physicists do, the chemists also like to give names to their toys, and this was called OLGA, *On-Line Gas Chromatography Apparatus*. In a mixture of reaction products and hydrogen gas, non-volatile elements like uranium (produced by proton transfer reactions from the target curium) were deposited on quartz powder in a 1,000 °C furnace. The volatile elements (112 to 116) were condensed onto cooled palladium foils mounted on a wheel. This wheel was rotated stepwise between pairs of detectors in order to count the deposited activity for α particles and spontaneous fission fragments.

If the superheavies were to be radon-like with noble gas properties, they would not stick to the palladium foils. To search for those species, the wheel was replaced by a cryogenic system. At temperatures as low as −200°C, even the gaseous superheavies would stick to the trap, a solar cell which simultaneously worked as a detector for fission fragments.

Although the sketch shown in Figure 10.9 looks simple, a lot of

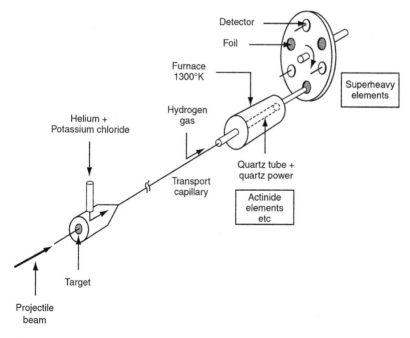

Figure 10.9 High temperature on-line gas phase chemistry used in search experiments for superheavy elements. The figure was taken from Herrmann (1988).

technical equipment is needed to keep everything running for many days of irradiation and everything has to be shielded because of radioactivity. Not that we had that much of it: our largest quantity was the milligram of target material, whereas in nuclear reactors several tons of radioactive material have to be shielded and controlled.

In as labour-intensive an enterprise as the Ca + Cm experiment, it would have been inexcusable not to look for long lived species as well. Some samples were flown to the Institute for Nuclear Chemistry of the University of Mainz where Günther Herrmann, Norbert Trautmann and their colleagues had built a low background counting system. This was located underground, well-shielded from cosmic rays. The detectors could measure single and coincident fission fragments and neutrons emitted during fission. The samples from Berkeley were measured for nearly two years to extend the half-life range covered to several years.

The result after all these efforts was frustrating: no significant α-event or spontaneous fission decay was found that might have originated from a superheavy nucleus.

The physics experiments started a few days after our arrival at Berkeley. At the time we had not yet heard in full about the negative results of our chemistry colleagues so we were still optimistic about making and finding the superheavies. The experimental arrangements were well prepared in advance by Ghiorso and his colleagues. Some members of the group, Matti Leino from Finland, Saburo Yashita from Japan, Jean-Pierre Dufour from France and Peter Lemmertz from GSI, had already spent some time at the Berkeley lab, while Peter Armbruster had arrived four weeks earlier. Gottfried and I came after the others because we still had to work on the analysis of our element 109 experiment at GSI which did not finish until September 2nd. When we all finally gathered in Berkeley, a journalist from *Discover* took a group photo of us in Ghiorso's cave (Figure 10.10).

The main experimental instrument in Berkeley was the *Small Angle Separator System* (SASSY), developed by Ghiorso and his co-workers over the previous few years. It was a gas-filled separator of the type already described in Chapter 9. An important difference between the various gas-filled separators is the deflection angle; at SASSY it was only 23° while the new (1999) Berkeley Gas-filled Separator has a deflection angle of 70°. At large deflection angles, separation of the reaction products from the beam particles becomes better but it also becomes more difficult to focus the fusion products onto the detector, especially if the electric charge state of the ions is uncertain.

Sometimes the deflection angle of an instrument is determined by such trivial things as concrete walls, and so it was with SASSY. It had to fit into a

Figure 10.10 The German–American element hunting team in Ghiorso's cave at LBNL in Berkeley. From left, standing Mike Nitschke, Matti Leino, Kenton Moody, Peter Lemmertz, the author, Saburo Yashita, Gottfried Münzenberg, Jean-Pierre Dufour, seated Albert Ghiorso (left) and Peter Armbruster. The figure was taken from *Discover*, December 1982.

relatively small cave; Ghiorso was very proud when he told to us that he had managed to build SASSY only because he removed part of the concrete walls working with a hammer and chisel. I admired him for doing that because it reminded me of our own work setting up SHIP. Experimental physics sometimes really means physical experiments.

The detector was a reconstruction of the system we used at GSI, position-sensitive silicon detectors. They were manufactured at the lab and were ready to be mounted in the focal plane of the separator when we arrived. Early on, there were some problems with position resolution but we easily solved them by changing the pulse shape of the signals.

It was interesting for us to visit the place where Kenton Moody prepared the curium targets and where other actinide elements were stored. It was a small room, glass cabinets covering the walls and behind the glass were numbers of little bottles and ampoules reminding me of the 'Phiole' in Goethe's *Faust*. (You may decide yourself whether Moody himself [Figure 10.10] resembled Faust.) The contents were certainly much more valuable than the amber room in St Petersburg. Some of the material was available nowhere else in the world and, even there, only in milligram quantities. It was the most valuable room I had ever ever been in!

The SuperHILAC delivered a good and stable beam of ^{48}Ca ions. The experiment ran smoothly except for one mishap on a Friday afternoon: the lab's power supply was switched off by the electricity company. They had permission to do so by a contract arrangement to lower costs for the lab. But the result for us was failure of the vacuum pumps and, instead of leaving Ghiorso alone so that he could concentrate on coping with the power failure, too many people started talking about too many different things. Moreover, it was dark and in all this confusion nobody thought about switching off the helium flow for the gas-filled separator. The pressure increased until the thin entrance window and the curium target broke and a large part of SASSY was contaminated. Poor Ghiorso; he now had to spend the weekend cleaning up the mess. The rest of us felt guilty but left him to it; without special permission, we were not allowed to help with the clean-up.

Although that weekend was not very nice for us, there were other weekends when the experiment did run well; our experiments usually ran day and night, seven days a week. Then we used our free time to explore the fantastic Californian landscape. In some ways it reminded us of Europe but here everything, the mountains, the trees and the distances, were all ten times bigger.

After the experiment ended, the first analysis of the data was a disappointment: nothing was found which looked like element 116. But Berkeley's theoretician Wladyslaw Swiatecki, whose ideas had contributed so much to getting the whole search started, was far from discouraged and tried to comfort us. When some of us were calling the present effort 'the last search for the superheavies', Swiatecki smiled and said almost apologetically that there were other possibilities. He was right in general but in detail the story was very different.

'A better reaction for making stable superheavies would be ^{48}Ca + ^{254}Es or even better ^{48}Ca + ^{255}Es. So in a sense,' concluded Swiatecki, 'you might say that this is the last search for the superheavies with available – though exotic – targets and projectiles. But if you've not performed reactions like ^{48}Ca + ^{255}Es you really can't say that you've tried everything. As long as an interesting experiment is merely very difficult but not actually impossible, it's usually just a question of time before it is tried.'

Few months after the experiment in Berkeley a repetition followed in Darmstadt. This time Ghiorso had been our guest and we knew him to be an enthusiastic bird watcher. GSI is in the middle of a forest surrounded by meadows, a very peaceful area not far from industrial centres, an ideal region for him in the early mornings to get out his binoculars and start his bird watching. It was springtime when he was there and he always came back to the lab happy and we let slip that, just for him, we had scheduled our collaboration for March. The experiment ran for almost three weeks in February/March of 1983. The targets were imported from Berkeley; radioactive targets were not common in our laboratory and we had to develop special protocols for mounting the targets, etc. That was easier in Berkeley and in the meantime we had accumulated experience with plutonium irradiation.

Technically, the SHIP experiments ran well but choosing the beam energy was problem. Nobody knew what the optimum should be so we decided to try five different levels but then split the measuring time also into five shorter intervals so as not to stake everything on one set of conditions. Searching for the unknown is always a risk and this time fortune did not smile on us: no decay chain was observed which might be a candidate for element 116. The chemistry experiments had been repeated at GSI and they, too, were unsuccessful.

Looking back, I realise that this attempt was too early. The equipment was not sensitive enough and there were too many uncertainties about the settings for the separators. Nobody knew the charge state of element 116 ion when it leaves the target; indeed, that is still a problem. However, at the time, opinion generally was that reaching the superheavies meant by-passing several unstable fast fissioning nuclei so some sort of a jump into the unknown seemed essential. Perhaps that was justification enough for our trying to do so.

The big surprise: low fission of even elements

The next heavy element run after the 116 adventure was a year later: it was element 108. After observing α-decay for element 109, we not unreasonably hoped that at least an odd isotope of element 108 would also undergo α-decay. The theoretical predictions, however, were very uncertain. Some calculations gave fission half-lives as short as 0.1 milliseconds so α-decay would have no chance to compete. For the α half-life of $^{265}108$, the isotope we planned to look for, our estimate was 1–10 milliseconds. This value was fairly reliable, based as it was on the measured half-lives of the neighbouring isotopes of elements 107 and 109, $^{262}107$, $^{266}109$. If we found only fission, a correct assignment would be very difficult but if we observed α-decay, we could apply our correlation method and establish decay chains ending in known daughter nuclei. Unfortunately, in the expected decay chain of $^{265}108$ the daughter $^{261}106$ was not yet known, only the granddaughter $^{257}104$. It could well be that $^{261}106$ also decays by fission so we had to start with $^{54}Cr + {}^{208}Pb \rightarrow {}^{261}106 + 1n$ in order to identify it.

We planned this run for February/March, 1984. Both ^{208}Pb and ^{207}Pb were irradiated in an attempt to have a better chance of also finding the even–even nucleus ^{260}106, produced with the first target in the 2n emission channel and with the second in the 1n channel. Those reactions had earlier been looked at by the people at Dubna; despite the fact that they could measure only fission events, they had hints from the half-lives that the observed fission might be related to element 104 isotopes and not directly to 106. If correct, that would mean that element 106 must undergo α-decay.

In our experiment at the SHIP we could now directly search for α-particles and as a result we identified three new isotopes, ^{259}106, ^{260}106, and ^{261}106. The odd isotopes decayed only by α-emission with half-lives of 0.48 and 0.23 seconds, while the even–even isotope had a fission branching of 50% and a half-life of 3.6 milliseconds (see Figure 10.1); the partial fission half-life is twice as long at 7.2 milliseconds. This meant that the fission half-life of the element 106 isotope was similar to that of element 104 isotopes: it did not decrease rapidly as theory predicted. Why was that, and could we expect a similar trend up to element 108? The only way to find an answer was to do the experiment.

Element 108 – hassium

The choice of the reaction to make element 108 was clear: ^{58}Fe + ^{208}Pb → ^{266}108* but once again we were uncertain about the beam energy. As in the case of element 109, theory predicted that extrapush energy (explain in the next section) was important. Finally, we decided to use an energy, 5.02 MeV/nucleon, slightly above an energy necessary to bring projectile and target nucleus into contact.

We ran for 10 days in March, 1984. Everybody was tense: would we observe fission or α-decay? Eleven days later we had the answer: three beautiful α-decay chains. The α-decays of ^{261}106 and ^{257}104 showed up again so the parent must have been ^{265}108; the mean half-life was 1.8 milliseconds. The most important finding was that there was no fission and the partial fission half-lives for even elements did not break down at least as far as element 108.

So, in April/May 1986, we used the reaction ^{58}Fe + ^{207}Pb to search for ^{264}108. Only one chain was measured which started with the short half-life of 76 microseconds for the α-decay of ^{264}108. Again there was no fission event. In more recent experiments in the mid-1990s, one more α-decay and two fission events were assigned to the decay of this even–even isotope with a mean half-life value of 0.26 milliseconds and the fission branching 50%. These data fix a partial fission half-life of 0.52 milliseconds for ^{264}108, a value which agrees well with the results of the most recent theories.

Our letter to IUPAC and IUPAP proposing a name for element 108 read in part: 'Darmstadt was the former capital of the German state, Hesse. 'Hassia' was the Latin name of the state in the middle ages. Our laboratory was founded in 1969 by the initiative of physicists and chemists from Hesse ... So we intended to follow an old tradition of naming an element after the location of its discovery. Element 108 should be named *hassium* (Hs).'

Two more neutron-rich isotopes, ^{267}Hs and ^{269}Hs, were later identified as daughter and granddaughter nuclei in the decay chains of ^{271}110 and ^{277}112, respectively. Shortly before our experiments on element 110 in 1994, three α-chains were observed in Dubna, confirming the discovery of ^{267}Hs using the reaction ^{34}S + ^{238}U.

Very recently, in May 2001, the chemical properties of hassium were investigated by a team of nuclear chemists from Germany, Switzerland, USA, Russia and China. At the UNILAC an intensive beam of ^{26}Mg was directed on a ^{248}Cm target. From the compound nucleus ^{274}Hs a number of 4 and 5 neutrons were emitted, thus creating the new isotope ^{270}Hs and ^{269}Hs (the granddaughter of ^{277}112). The proper place in the periodic table was obtained from the formation of HsO_4 (similar to the well known and poisonous compound OsO_4), which showed that hassium belongs to the transitional elements in group VIIIA just below osmium (where it ought to be).

The nightmare – extra-push

One subject brought up in the early 1970s led to endless, boring discussions in our lab: the concept of extra-push and extra-extra-push energy. The basic idea was that if nuclei do not want to fuse, they need an extra push and if they still don't want to, then OK, an extra-extra-push!

However, sometimes nuclei behave like people – or donkeys. If they are pushed, they rebel and even get angry. If they get angry enough, they might even explode and nobody knows what will happen then.

The extra-push concept came from theoreticians, from Wladyslaw Swiatecki and William Myers at Berkeley, some of the best people in the heavy ion reactions field. Their idea was simple and brilliant, although it buried any hope for producing superheavies.

If we regard the projectile and target nuclei as liquid drops (Figure 3.3), we can estimate the forces acting on them during the processes of fusion and its inverse, fission. Principally, these are Coulombic forces (i.e. static electric repulsion acting between and inside highly positively-charged nuclei) and nuclear attraction. It is the difference between the two forces which determines how the fusion or fission process goes on.

The Coulombic forces are weak but act at long range. We know this from their ability to keep the negatively charged electrons in orbits at a large distance around the nucleus. This is similar to gravitation which also acts up to large distances. The nuclear forces have a short range, not much more than the size of a nucleon, but are much stronger than the Coulombic forces. We can illustrate the way short-range attractive forces work from the behaviour of water drops or little spheres of mercury, although those forces are much weaker than the nuclear ones. If you bring two of these drops very close to one another, nothing happens. But let the surfaces touch, then – blop – the two drops coalesce into one. Because the shape of a drop is determined by the forces acting on its surface, this fraction of the short-range attractive forces is named 'surface forces'.

Imagine further that both drops are positively charged and that the charge increases with size. Small drops behave like uncharged drops, two little drops will – blop – join up when they touch each other. Bigger drops, however, will not coalesce so easily and above a certain size (or charge) they will not coalesce at all. If we try to bring them into contact, the electric forces will repel them. We can try to force the drops to coalesce by giving them more speed, an 'extra push'. If the speed is high enough, the Coulombic repulsion will be overcome although the fused drop will probably vibrate strongly and burst.

Myers and Swiatecki have quantified this phenomenon for nuclei; Figure 10.11 demonstrates what might happen. The figure is a set of energy curves for various products with nucleus charges from $Z = 20$ to $Z = 114$. The x axis shows the charge ratio of the projectile to the total charge of projectile plus target. At a value of 0.5 the system is symmetric but at 0.1 it is asymmetric, as indicated by the insets showing the projectile and target nuclei at the moment of contact. The curves represent the total energy (normalised at $x = 0$) of two spheres brought into contact and are a measure of the opposing effects of Coulomb repulsion and surface attraction.

For a product of Z that is less than 60, the slope of the energy curve is such that any set of projectile and target will be driven towards a spherical compound nucleus. However, for higher Zs and symmetric projectile and target combinations, the forces are such as to drive the system towards fission. Notice that, in this model, projectiles heavier than neon ($Z_p = 10$) are very unfavourable for Z greater than 100 (this is at $x = 0.1$). This warning was addressed to the planners of heavy ion accelerators at the beginning of the 1970s.

Over a couple of decades, theoreticians worked out models to obtain values for extra-push by undertaking hydrodynamic calculations of the collisions of charged viscous liquid drops with large surface tensions, also taking into account the internal structure of projectile and target nucleus consisting

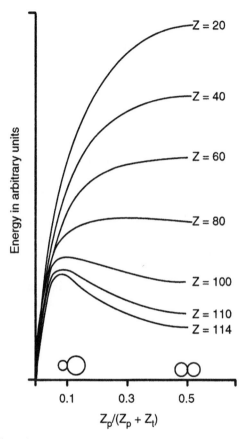

Figure 10.11 Effect of target-projectile charge asymmetry on fusion and re-separation of a system with total charge Z. See text for further explanation. The figure was taken from Crandall (1974).

of protons and neutrons. An increasing amount of extrapush was predicted for the fusion of heavier and heavier nuclei.

Despite the gloomy prospects we started our first element 110 experiment in March 1985, using the reaction $^{64}Ni + ^{208}Pb \rightarrow ^{272}110^*$. It was our 100th run on the SHIP which should have been a good omen. The beam energy chosen would have given an excitation energy of the compound nucleus of 20 MeV. Half-heartedly following the advice of theoreticians, we took into account an extrapush energy of 10 MeV in addition to the value of 10 MeV predicted by a simple fusion model which did not include the extra-push theory. This was a mistake as turned out later. After two weeks of irradiation the result was negative. This result was disillusioning, and not only for the synthesis of element 110. It might mean, as some of us were beginning to believe, that elements beyond 109 could not be made because the extra-push energies would be too high. That would be the end of cold fusion for the synthesis of new elements.

Our one remaining hope was the use of more asymmetric systems for which less or even no extra-push was predicted (Figure 10.11). We tried once more to synthesise element 110, now using the reaction $^{40}Ar + {}^{235}U \rightarrow {}^{275}110^*$. This is a so-called 'hot fusion' reaction (see Chapter 6) because, at barrier energy, the excitation energy of the compound system amounts to about 45 MeV compared with 10 to 20 MeV for cold fusion. Argon beams are fairly easy to make and uranium-fluoride (UF_4) is a target with a high melting point, so we could use high beam currents.

The irradiation ran for five days in 1986. With a large number of projectiles there was neither fission nor α-decay. Maybe it was the time to remember what we had said at the end of our 109 paper: 'Our hopes of making still heavier elements, such as element 111 by using ^{64}Ni projectiles, are small since we have obviously reached the limit of production cross-section that can be observed using present-day techniques.' Perhaps we had. Something needed improving.

11

The year 1988: a new beginning

It was 6:30 on the morning of Sunday February 14th, 1988 and an experiment designed to repeat and confirm our element 109 findings had just ended. All physical results should be reproducible so we felt we should do so – and again we were successful. Only one atom had been observed in the 1982 experiment and now we found another. It took a load off our minds. Now, closing the door on element 109 and the lighter elements, a new era of experiments could begin. Simultaneously, major changes were taking place at GSI as people left and were replaced, and many of the experimental arrangements changed too.

The old SHIP group came together for a last time in 1988 to do the second 109 experiment: Peter Armbruster, our department leader, Gottfried, the designer of the SHIP separator, Willi, Karl-Heinz and Hans-Joachim, several guests and students, and I.

Armbruster left GSI in October 1989 for three years to be Director of the Laue Langevin Institute in Grenoble. Would it be possible to assemble and retain a new group of enthusiasts? Even more uncertain, would it be possible to continue the experiments and find a new element? Both difficult questions, but we were confident and had plans.

Gottfried became the acting head of Nuclear Chemistry II during Armbruster's absence; as an expert in the design of separators he started to build a new one for nuclear fragments produced in high energy collisions. His new facility gained renown as 'FRS', the *Fragment Recoil Separator*. Willi left for high energy physics and joined a group looking at the behaviour of nuclear matter in high energy nuclear collisions. Karl-Heinz began to prepare for experiments with the FRS. And I became leader of the SHIP group; thanks to Gottfried, I moved up a grade in the hierarchy. My group now comprised the young physicists Fritz Heßberger and Victor Ninov, our electronics engineer Hans-Joachim Schött and our technician Hans-Georg Burkhard. I could count on Gottfried and I had the telephone number of Armbruster in case we needed expert and urgent advice. I felt free – free to make decisions, free to choose equipment and free chart my own scientific way forward.

An important first step was to complete the crew, something I was able to do during a conference at Dubna. Heavy element research had a long tradition in Dubna. The advantages of recoil separators were realised and at the time they had completed an instrument similar to SHIP. It was called VASSILISSA after a character in a Russian fairy tale. We had problems in common with them and could help each other with experimental experience, material and resources. The end of the Iron Curtain in 1989 greatly helped our unconventional start into a joint venture. We did not want to make plans for the following five years which would probably not have been fulfilled but would have meant a lot of paperwork. Instead, we planted a seed, watered it from time to time (with some vodka there and some red wine here, I admit) and let the plant grow. If the soil was rich, the plant would grow by itself and that is actually what happened.

We also tried to reinforce our contacts with colleagues at Berkeley. However, at the time the Berkeley people were not interested in collaboration and so we could not recruit any of the young or older scientists from there. However, some years later in 1997, Victor Ninov was hired by the Berkeley people. He started to build a new gas-filled separator at the 88-inch cyclotron, the BGS which stands for *Berkeley Gas-filled Separator* and which earned its merited reputation in 1999.

Several of the Dubna people, as well as colleagues from Slovakia and Finland, regularly came to GSI to co-operate in the experiments. Our crew was complete.

Whether we could achieve our targets was uncertain; the main hints were theoretical predictions. Our best theoreticians had for years been working to calculate the properties and co-ordinates of the postulated superheavies but their predictions were, like a map of Treasure Island washed up onto the shore in a bottle, damaged and difficult to read after years in the sea.

In 1988 our UNILAC was fitted with a post-accelerator. This was a synchrotron, a ring accelerator, able to accelerate a bunch of ions up to a speed 80% of the velocity of light. The synchrotron was called the *Schwer-Ionen-Synchrotron (SIS)*, ('Heavy ion synchrotron'). Fortunately this was not the end of our heavy element research which needs low beam energies, ions travelling at only about 10% of the velocity of light – quite the contrary!

Synchrotrons need time for the acceleration. When a pulse of ions is injected, it takes them about a second to get up to speed. There is enough time between these pulses to perform low energy experiments using the old UNILAC in its original format of 20 years earlier. We also built a second ion source: there were then two independent ion sources so the high energy people could work with beam particles different from those we needed for heavy element production.

Inventory

Conferences, in general, are places to float new ideas or new data. That is easy if you have new data; if not, you can always offer a review or present a summary! That is what I did in 1988, in Zakopane in April and in Dubna in October; I chose to review. My main motivation was to order my thoughts and present a clear picture of what was actually known, what was possibly known and what was certainly not known. Then I would decide how to continue our experiments at the SHIP.

The elements up to $Z = 109$ were known. Their lifetimes decreased continuously down to milliseconds (thousandths of a second): although milliseconds are extremely short in normal life, in elementary processes that time span is almost infinitely long. Our detection limit was a microsecond (a millionth of a second), 1,000 times shorter still. The newly calculated lifetimes made us quite optimistic that isotopes of elements 110 and heavier would live long enough to be observable.

Calculations of half-lives and decay modes had made great progress in the mid-1980s, when a reassessment of the theoretical model became necessary because of newly-measured decay data. The most important experimental result was that the likelihood of fission was fairly constant and did not increase. The measured data on α-decay up to element 109 allowed us to determine the binding energies of the decaying nuclei. Once having measured that, we could estimate the influence of the nuclear shell structure on the binding energy. To do that we assumed that the nuclear binding energy is composed of a major fraction originating from the liquid drop properties of a nucleus (the macroscopic part) and a minor fraction due to the shell structure (*shell effect* or microscopic part). Because the binding energy is negative, a negative shell effect results in more strongly-bound nuclei. The macroscopic part could easily be calculated, and subtracting it from the measured binding energy gives the shell effect. To our surprise, we observed an almost regular decrease of the negative value for the shell effect which meant that, up to $Z = 109$ and $N = 157$, the shell effect tends to stabilise the nuclei.

That was the experimental result; the theoreticians had to find the reason. Actually, an explanation had already emerged in some of the publications from the late 1960s but at that time everybody was fascinated by the double magic superheavies at $Z = 114$ and $N = 184$, so that another minimum of the shell effect at $Z = 108$ and $N = 162$ was overlooked. This second minimum we had now rediscovered experimentally.

Although ^{266}Mt is five neutrons away from $N = 162$, it already benefits from the shell effect and the new calculations revealed a whole region of shell-stabilised nuclei around $Z = 108$ and $N = 162$. However, in contrast to the spherical superheavy nuclei, these nuclei are deformed. The theoreticians thought they would look like barrels.

The measured half-lives correlated well with the new calculations and we could therefore trust our ability to predict half-lives for the unknown species but this was not the case for the cross-sections. An overview of measured cross-sections at the time is given in Figure 11.1. The three panels show the trend for increasing element number for hot fusion (4 neutron evaporation channel) and cold fusion (2n and 1n channel). The data points follow straight lines on the logarithmic plot but with different slopes. The lines were extrapolated in order to estimate the cross-section for the synthesis of element 110. The result was that the highest cross-section was obtained for cold fusion and the one neutron channel, 1.5 picobarn. One problem was solved.

The remaining problem was the question of optimum beam energy. Prediction indicated the need for an extra-push to fuse heavy elements using lead targets but the amount was unclear. Experimental data offered no help because there were too few data. In this situation we had to measure excitation functions, the cross-section as function of the beam energy. Only then

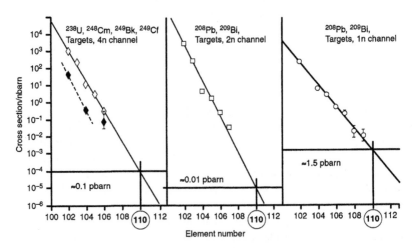

Figure 11.1 Extrapolations of experimental cross-sections for the preparation of a new experiment to search for element 110. The data are cross-sections of residues after fusion reactions and evaporation of four, two and one neutron (from the left). Targets of ^{249}Bk, ^{249}Cf (Berkeley data), ^{238}U, ^{248}Cm (Dubna data, full symbols) and ^{208}Pb, ^{209}Bi (Darmstadt data) were irradiated with light projectiles of similar proton to neutron ratio. The favourite reaction for the production of element 110 is ^{62}Ni + ^{208}Pb $\rightarrow$ ^{269}110 + 1n, with a cross-section of about 1.5 picobarn.

the maximum cross-section could be determined and, most important, at which beam energy. This excitation function was needed for at least two known elements below 110. Then, assuming a linear trend of the maximum, extrapolation would give the optimum value of the beam energy for production of element 110.

We decided to measure the excitation functions for elements 104 and 108 using the reactions $^{50}Ti + ^{208}Pb \rightarrow ^{258}Rf^*$ and $^{58}Fe + ^{208}Pb \rightarrow ^{266}Hs^*$. The following reasons were decisive:

1. element 104 had a high cross-section and within one to two weeks we could measure excitation functions very accurately;
2. element 108 has a small cross-section but is the even element neighbour of element 110. The location of the element 108 cross-section maximum would be of major help in determining the beam energy for element 110.
3. the series of projectiles ^{50}Ti, ^{54}Cr, ^{58}Fe and ^{62}Ni on ^{208}Pb targets results in isotopes of element 104, 106, 108 and 110 which differ by only an α-particle (see Figure 10.5). We could therefore expect the smallest changes in the reaction mechanism and a reliable identification of element 110 because the daughter nuclei after α decay were already known. Improved decay data on the daughter nucleus ^{265}Hs could also be expected from the measurement of the excitation function.

So to the planning. How long a beam time would we need? The estimate was easy to make, the result was sobering. The three events of ^{265}Hs had been measured within 12 days. For an excitation function we would need 10 times more, i.e. 120 days of beam time. The limit which we reached for the synthesis of element 110 in 1985 using $^{64}Ni + ^{208}Pb$ was obtained within 15 days. To push that limit down to 1.5 picobarns results in 150 days of beam time. Both experiments add up to a total of 270 days, almost a year, of pure beam time, 24 hours a day, 7 days a week, no pause, no holidays. Nobody wanted to commit to such long beam times; the only way out would be to increase the experimental sensitivity.

That took a lot of effort and lasted five years. But we were not constantly working on the project; there were delivery times, beam times for other investigations, conferences, paper work and a host of other activities in between. Many of the changes were highly technical but few are easy to appreciate. It is interesting to look at photos of the instrumentation which was installed at the time and is still in use.

We had to have a new target wheel and asked Helmut Folger whether his group could prepare longer lead targets so that each would move through the beam within a pulse length of 6 milliseconds. The targets had to be as high as possible in order to defocus the beam and reduce the thermal

Figure 11.2 The new 1994 target wheel, rotating at 1,125 revolutions/minute through the beam, carried eight large area lead or bismuth targets. The duration of the beam pulse could be extended up to six milliseconds. The outer diameter of the wheel was 356 mm and 310 mm at the centre of the targets. The photo shows irradiated targets, and traces of the beam are visible.

heating, enabling us to use higher currents. As in the original design, the target frames could not be hit by the beam halo or the background radiation would increase. The new wheel is shown in Figure 11.2. The synchronisation stability of the wheel drive was also improved. One major benefit was that we no longer needed the night shifts and were not afraid of long beam times.

A major improvement was a change of the target position closer to the SHIP. A lot of mechanical work was needed but the result was a factor of two higher transmission because of the larger solid angle available. Now, 50% of the produced nuclei would reach the detector.

Detectors and electronic devices for signal processing and data acquisition were reconstructed. The detectors needed to be larger and more sensitive. We could not risk missing any of those rare events which were produced with so much effort. Figure 11.3 shows two of the three time-of-flight detectors the ions have to pass before they are stopped in the silicon detector; the latter is

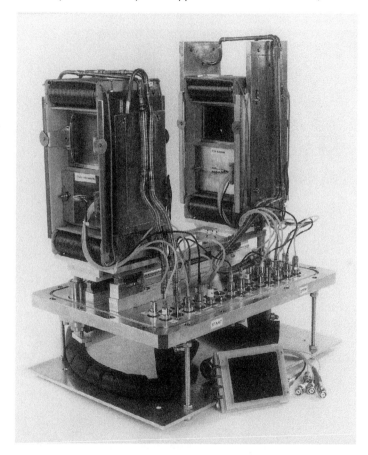

Figure 11.3 Photograph showing two of the three time-of-flight detectors usually used. The detectors are mounted on a vacuum flange ready to be fixed to the vacuum chamber. All particles escaping SHIP have to pass thin carbon foils of the detectors before they are implanted into the silicon detector. The size of the foils is decreasing down the beam, the entrance foil of the second detector is removed. When passing through the foils, electrons are emitted and accelerated by an electric field. Each detector has two coils made of a copper wire which create a magnetic field for bending the electrons to the surface of channel plates which act as electron multipliers. A stack of two channel plates is shown at the bottom of the photograph.

shown in Figure 11.4. Alpha particles or fission fragments escaping from this detector are stopped in a detector box (Figure 11.5) which was mounted in front of the stop detector. We had to leave just the hole in the centre open so that the ions could enter. Behind the stop detector were mounted germanium detectors, sensitive to X- and γ-rays. The sketch in Figure 11.6 shows how the

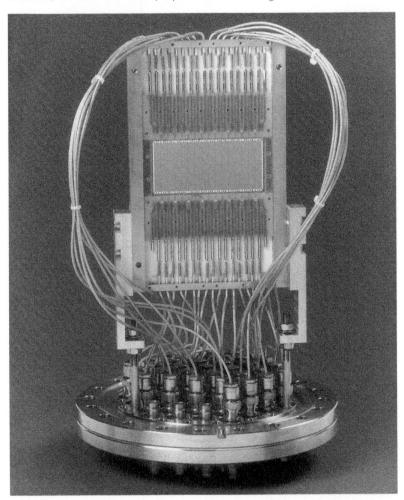

Figure 11.4 Photograph of the new large area silicon stop detector (wafer area = 80 mm × 40 mm). It is mounted on a vacuum flange with the cover removed to show the contacts to the top and bottom electrodes. The wafer is subdivided into 16 strips which are position sensitive in the vertical direction.

Figure 11.5 The detector box measuring escaping α-particles mounted inside the vacuum chamber; the way it fits into the array of detectors is shown in Figure 11.6. The stop detector, shown in Figure 11.4, is fixed to the back of the detector box. The array of silicon detectors is cooled from outside by a liquid to −15 °C via the flexible tubes near the bottom.

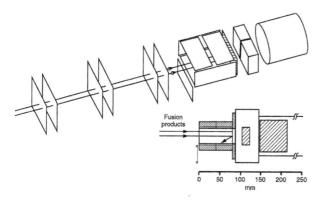

Figure 11.6 Assembly of detectors for the identification of new elements. They comprise large secondary-electron time-of-flight detectors (shown in Figure 11.3), position-sensitive silicon strip detectors (shown in Figures 11.4 and 11.5) and germanium detectors.

Figure 11.7 The detector area behind the SHIP. The photograph shows the 7.5°
deflection magnet, detector chamber, Dewar flasks filled with liquid nitrogen for cooling the germanium detectors and the cooling aggregates for
bringing the silicon detectors to −15 °C. The stop detector is mounted at
the top with the backward detectors hidden behind the vacuum chamber.
The two circular instruments on top of the chamber measure the temperature of the cryogenic pumps which keep the chamber on a pressure level
of 10^{-7} torr. The shaft for the degrader foils is visible just in front of the
two instruments.

detectors are put together, while Figure 11.7 pictures the detector area behind
SHIP. With this arrangement we could measure 100% of the separated ions
and of their α-decay or fission and, most importantly, down to lifetimes of
three microseconds.

The electronics for processing the detector signals was installed in a
specially built air-conditioned room located close to the detector area in

Figure 11.8 The electronic room for control of the experiment, signal processing and data acquisition.

order to keep the cable lengths short and to avoid electric disturbances (Figure 11.8).

In any exploration of the unknown, working out the methods is often the most laborious part of the investigation. We had learned that during the previous five years; now was the time to find what awaited us after all that effort.

12

Three more elements: 110, 111 and 112

Valuable preparation

By summer 1994 we were ready to take our first step towards element 110 and the excitation function of element 104. For our 12-day run the reaction used was ^{50}Ti + ^{208}Pb → ^{258}104*, increasing and decreasing the energy as required. The result was a yield curve as a function of the beam energy. We measured one curve with a well defined maximum for the evaporation of one neutron (this results in the isotope ^{257}104), a second curve at higher energy for the evaporation of two neutrons and, at a yet higher beam energy, a third for the evaporation of three neutrons (Figure 12.1). The last, however,

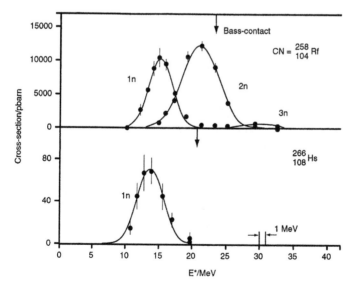

Figure 12.1 Excitation functions for the synthesis of isotopes of elements 104 (rutherfordium) and 108 (hassium). The reactions used were ^{50}Ti + ^{208}Pb → ^{258}Rf* and ^{58}Fe + ^{208}Pb → ^{266}Hs*, respectively. The excitation energy E* is proportional to the beam energy plus a constant value of (negative) binding energy of the reacting nuclei.

showed a strongly decreased yield as a result of increasing fission of the compound nucleus. It was a good start.

The actual breakthrough came with the next series of experiments. It started with the excitation function for element 108 and was our longest irradiation period, lasting from October 6th to December 18th and was our most exciting one. A few years ago the excitation function of element 108 would have been almost impossible to measure because the cost of enriched ^{58}Fe needed to prepare an intensive beam had increased to about $500,000 per gram; for the experiment we thought we might need about four grams. We were fortunate to get help from the Dubna people: when they arrived in Darmstadt they had the world's most expensive chunk of iron in their bags! It was not packed in a leather suitcase and there were no body guards around; the treasure arrived in a glass and a plastic ampoule wrapped in a Russian newspaper. There were five larger and 12 smaller pieces, totalling 20.7 grams comprising 93% ^{58}Fe.

We put the pieces on my desk and took a photo (Figure 12.2). About four grams went to the people running the ion source; the rest was in a vacuum flask. The Russian iron worked well in our ECR source: the beam intensity was high and only about 4 milligrams were consumed each hour. The ^{58}Fe beam ran for 29 days in October/November, 1994. The atmosphere was tense as we wondered how many atoms we would add to the three observed 10 years earlier.

The first week brought only one more; that was frustrating. How could we make sensible measurements if the counting rate continued like that or even declined? Should we stick with our original choice of beam energy? (This was the data point at 20 MeV shown in the lower part of Figure 12.1. It was at the same energy as in our 1984 experiment). Views were diametrically opposed. Armbruster, now back from Grenoble, was still in favour of the extra-push theory and wanted to increase it but the impression I had from all the cold fusion data was a trend to lower excitation energies. We could not agree. Finally, on Friday, October 14 at 5:30 p.m., I wrote in the log 'Energie-Wechsel von 5.02 nach 4.96 MeV/Nukleon' (*Energy changed from 5.02 to 4.96 MeV/nucleon*) and asked the operators to use the lower energy after the next change of the ion source. Would the cross-section increase? The tension mounted but we tried to be patient, perhaps having to chew our fingernails for another week before getting a result.

Fortunately, we had to wait for only a few hours. The ECR-ion source was charged with a new 450 mg piece of ^{58}Fe and the beam again went on target at 04:00, October 15th. The first event came at 22:46 that evening and the second only an hour and a half later. When, after four days, we stopped

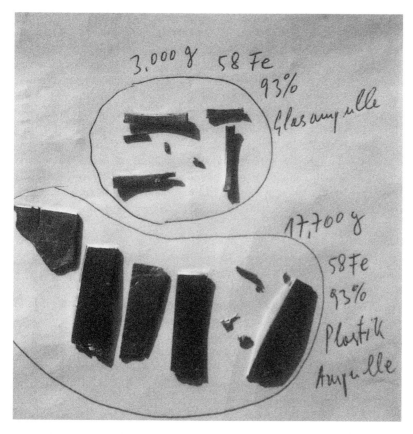

Figure 12.2 The million dollar treasure: 20.7 grams of highly enriched ^{58}Fe from Dubna.

the irradiation at that energy level, we had a total of 12 events, and the cross-section was ten times higher than in the previous irradiation.

Things were brightening up and the way forward had become clear; no extra-push energy was necessary. We made bets about how large the maximum production rate would be if we further decreased the energy. Estimates varied and we decided to measure in Deutschmarks the difference between each guess and what the actual value turned out to be. Everybody had to contribute that sum towards a dinner in a Thai restaurant. Eventually, with DM 300 in the kitty, we had quite a feast.

We knew that the Dubna–Livermore collaboration had started their 110 experiment on September 10th using the hot fusion reaction

^{32}S + ^{244}Pu → ^{276}110* and the gas-filled separator. There might an announcement any day that they had made element 110 but we had to keep cool enough to measure our excitation curve as well as possible. The complete curve is shown in the lower part of Figure 12.1. We got a total of 75 decay chains of ^{265}108. The yield for the evaporation of two neutrons had almost disappeared in the case of element 108 due to the increased fission of the compound nucleus. One more result from the data was the α-energy spectrum, which later turned out to be important for the identification of ^{269}110.

From our 75 decays of ^{265}108 we obtained two clear and well-separated α-lines. The two events from 1984 (the third one had been an escape α particle) agreed with the more intense lower energy line. However, as was found later, the ^{269}110 decays went by the higher energy α-decay so the two α-lines of the ^{265}108 decay had to be assigned to different energy states.

Element 110

On the basis of the new data, the search for element 110 was now easy. Our reaction was ^{62}Ni + ^{208}Pb → ^{270}110*. It began the day we finished with the 108 experiment and, after only two days, we found it.

The irradiation started on Monday. The beam had worked well from the beginning. On Wednesday, Gottfried invited our guests to be tourists for a bit. Next morning they were to leave for a trip to the Rhine. That Thursday I arrived in the institute at around 10 a.m. and, before I could drink my coffee, Victor gave me a hug and a slap on the back and said: 'We've got it!' The tapes were usually changed daily and the off-line analysis was carried out by Fritz and Victor during night; in the morning the file could be examined on the terminal screen.

That morning they found an interesting event and the three of us sat down to discuss it. It soon became clear that it must be the decay of ^{269}110, showing α-decay 393 microseconds after implantation in the detector. The α-energies for the daughter ^{265}108 and the granddaughter ^{261}106 were also measured but, unfortunately, that of the ^{257}104 decay got away. The decay chain is shown in Figure 12.3.

At about noon that very day it was clear that we had to publish our result as soon as possible because of the experiment running at Dubna; we thought a Short Note in *Zeitschrift für Physik,* in which we had published our earlier discoveries, would be most appropriate. We agreed not to delay by telling anybody about the discovery and to finish the draft paper by that evening. It was just as well that, apart from Armbruster, our colleagues were away but he did come in from time to time to discuss something or other, which held things up and made us nervous. We had really wanted to surprise him with a completed paper on his desk next morning.

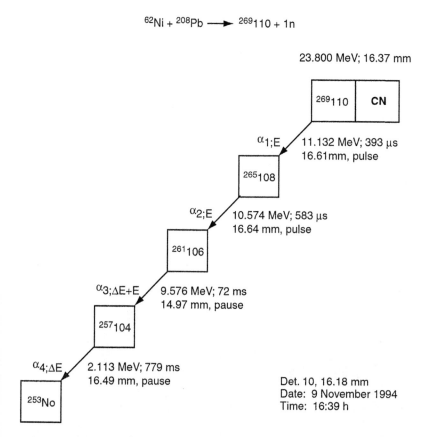

$$^{62}Ni + {}^{208}Pb \longrightarrow {}^{269}110 + 1n$$

23.800 MeV; 16.37 mm

| ^{269}110 | **CN** |

$\alpha_1;E$ 11.132 MeV; 393 µs
16.61mm, pulse

^{265}108

$\alpha_2;E$ 10.574 MeV; 583 µs
16.64 mm, pulse

^{261}106

$\alpha_3;\Delta E+E$ 9.576 MeV; 72 ms
14.97 mm, pause

^{257}104

$\alpha_4;\Delta E$ 2.113 MeV; 779 ms
16.49 mm, pause

Det. 10, 16.18 mm
Date: 9 November 1994
Time: 16:39 h

^{253}No

Figure 12.3 Decay chain starting from element 110. It was measured on November 9, 1994, at 4:39 p.m. The isotope ^{269}110 was created after evaporation of one neutron from the compound nucleus (CN). After separation by SHIP it was implanted into detector strip number 10 at a position of 16.18 mm from the bottom.

All three of us will forever remember that day. Fritz and Victor prepared the numbers and the figures while I wrote the text. Spotting our unusual behaviour, Armbruster might have sensed that something was up because he came into the office ever more frequently as the evening wore on. At one time he made the useful comment that Short Notes could now be four pages long instead of only two, as had earlier been the rule. This helped a lot because we would have room to mention the preparatory experiments but it also meant that we needed more time to write it.

Later, our friends came back from their Rhine tour but they were tired and, when they heard that everything was running well, went to bed early. We finished at 3:00 a.m., made copies of the draft, put a copy out for every member of the group with a request for comments and corrections – and went to bed. I had never before written a paper so fast.

Next Monday, November 14th, the final version of the paper was finished and one of our drivers, who happened to be going to Heidelberg, delivered it directly to the editors. We also officially informed the GSI staff and a press release was issued on the 17th; the following day we read about our results in the newspapers. There was also a group photo (Figure 12.4). Finding the new element was reason enough to be happy but our burden had become heavier because the experiment was still running (we wanted to find at least one more chain) and we had to satisfy a lot of interested people – all our colleagues at GSI and those from other laboratories around the world (extensively using e-mail) and journalists phoning in.

Figure 12.4 The discoverers of element 110. From the left: Alexander Yeremin, Andre Popeko, Viktor Ninov (front), Fritz-Peter Heßberger, Andrei Andreyev (rear), Matti Leino, the author, Gottfried Münzenberg (rear), Peter Armbruster, Jürgen Bossler (rear), Helmut Folger, Ursula Vogel, Hans-Joachim Schött (rear), Stefan Saro and Hans-Georg Burkhard.

The irradiation ended on the 20th after 12 days; the beam energy had not once been altered. Three more decay chains were found, the two longest chains of which could be followed down to ^{249}Fm. Having taken the precaution of leaving some empty space at the end of our paper, we appended the additional information as a note added in proof. There was no news from Dubna!

The second isotope – ^{271}110

It was known that cross-sections decreased when lighter isotopes were used as projectiles or targets so we had reason to hope that it would increase in the opposite direction, changing the beam from ^{62}Ni to ^{64}Ni. Meanwhile, we had learned how to estimate the optimum beam energy so that projectile and target nuclei came to rest on contact. The necessary energy is easy to calculate, because in cold fusion the reaction partners are spheres and, at that large distance, it is sufficient to consider only Coulomb forces and the well known nuclear radii and surface diffuseness. After three and a half days we had collected six decay chains of the new isotope ^{271}110. There was just enough time to try another energy step.

Our beam time officially ended November 30th and the last experiment of the year – a further investigation of the 'electron–positron creation' in uranium plus uranium collisions – was due to terminate on December 22nd. (Remember the predicted electron–positron pair creation in strong electric fields discussed in Chapter 7). So far, the outcome from those experiments was puzzling: sometimes positron lines were measured, sometimes not and that experiment is certainly worth another story. Now we were competitors: everything was going so well late in November that it would have been a pity to stop and so we suggested to our director that we should continue with the SHIP experiments until December 22 and shift the 'positron experiment' into 1995 because we wanted to use the time to search for element 111. Although not the unanimous view of our group, it certainly got the most votes.

The director agreed and, with the prospect for element 111 in mind, we increased the beam energy slightly in the 110 experiment then still running. While it was clear that we would use ^{64}Ni searching for 111 as well as for 110 (we had only to replace ^{208}Pb by ^{209}Bi), the optimum beam energy was not clear. So, in our remaining time, we looked for an excitation function for element 110 which might help us to choose the best beam energy in the element 111 search; the excitation functions for the production of odd heavy elements were not well enough known to estimate the beam energy for the synthesis of element 111.

At the third and highest beam energy, we measured only one more chain of $^{271}110$ in 4.8 days. It was clear now that the cross-section did not increase at higher beam energy. On November 30 we switched to element 111.

Element 111

The average decrease of cross-sections in cold fusion is 3.6 per element so we could make a guess for element 111. The time remaining before Christmas might be enough to find it.

On December 1, 1994, we started an irradiation of ^{209}Bi, retaining the ^{64}Ni beam from the $^{271}110$ experiment. After five days with no event, we increased the beam energy and at 05:49 on December 8, we registered the first decay chain but, for technical reasons, it seemed to us too weak a piece of evidence on which to claim the discovery of the new element 111.

On the sixth day we changed the energy again and within another six days found two more chains, the longest (and therefore the most significant) of which went down to the decay of ^{256}Lr (see Figure 10.1). That was sufficient for claiming element 111.

After 17 days, we stopped the ^{209}Bi irradiation on December 18th and tried to complete our data collection for the ^{64}Ni + ^{208}Pb reaction in the remaining four days of our beam time allocation. Our paper with the title 'The new element 111' went, as usual, to the *Zeitschrift für Physik*; it arrived before Christmas.

A beam time of 77 days was behind us. If you enjoy numbers, you might like to know that in all we had irradiated targets of ^{208}Pb and ^{209}Bi with 4.4×10^{18} ^{58}Fe ions, 2.2×10^{18} ^{62}Ni ions, as well as 2.1 and 3.2×10^{18} ^{64}Ni ions for elements 110 and 111, respectively. That total of 5.3×10^{18} ^{64}Ni ions amounts to the weighable quantity of 0.56 milligram. We had received extremely important new information on the reaction mechanism for the synthesis of heavy nuclei by cold fusion, rounded out our view of the stability of isotopes near the neutron number $N = 162$ and could synthesise and unambiguously identify two new elements. Not bad going and, we felt, we had given ourselves a good Christmas present.

Element 112

For the element 112 search we again used the proven 1n cold fusion channel. Our experience with ^{62}Ni and ^{64}Ni gave us encouragement that the cross-section for element 112 synthesis using the reaction ^{70}Zn + ^{208}Pb → $^{278}112^*$ would not follow the average cross-section decrease. We were optimistic that we were in with a chance. Over 24 days in January/February, 1996, using only one beam energy, we found the first decay chain on February 1st and a second on the 9th. Both were assigned to $^{277}112$. Easy, wasn't it?

What had we learned so far?

We had learned that the pessimists were wrong. That does not mean that the optimists were right because there were no optimists: just a few people who thought that we might think about our results when we had done some reliable and sensitive experiments. So, the first topic of the three we promoted was how to perform those reliable and sensitive experiments: on that basis we were able to determine the stability of heavy elements and how to produce them. At that time no more than a few nuclei up to 112 were known and only few reactions tested. Let us look first at the interim results.

The α-decay chain of element 112 was final proof that fission is not the dominant decay mode of heavy elements. It does not mean those heavy nuclei do not fission at all but that the partial fission half-life is much longer than the partial α-half-life. Thus, α-decay dominates. The reason for the decreased fission probability is an increased binding of the nuclei in the region of nuclei near $Z = 108$–110 and $N = 162$. The increased binding was theoretically predicted and then proved by the decay chain of $^{277}112$. Figure 12.5 (bottom) shows the decay chain across the map of nuclear binding energy. Only that part of the total binding energy due to the nuclear shell structure ('shell correction energy') is plotted. This energy has negative values in the case of increased binding. Figure 12.5 clearly shows the minimum at $Z = 108$–110 and $N = 162$, the decay of $^{277}112$ just going straight through it.

The structure in the upper part of the figure looks like a constellation of stars but actually is the measured α-half-lives of the decaying nuclei. It begins on the lower right side with the decay of $^{277}112$ and $^{273}110$, both with values of a few hundred microseconds. Then there is a big jump to ten seconds for the decay of $^{269}108$, the decays continuing down to nobelium. The explanation for this structure is given by the α-decay energy which is high for decays in the direction of a shell correction energy minimum, resulting in short half-lives. The decay energy is low for decays away from the minimum and the half-lives are longer. All these properties were measured.

Theory predicts another property of nuclei near this minimum: they should be deformed, looking like barrels. In principle it is possible to determine the deformation experimentally and, although this has not yet been done, we have no doubts that theory correctly describes this property.

People sometimes argue that, from the shell correction energy, the nuclei around this deformed minimum should already be called 'superheavy'. However, at $Z = 108$ and $N = 162$, the proton and neutron shell is not closed and the nuclei are not spherical, so I prefer to reserve the name *superheavy* for those nuclei having closed shells.

Where should we search for those double magic spherical superheavies? Theory had accurately predicted the location and properties of deformed

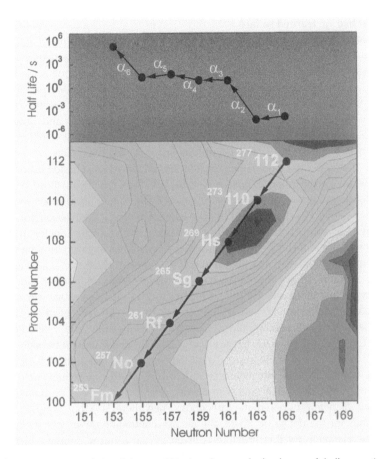

Figure 12.5 Decay chain of element 112 plotted across the landscape of shell-correction energy. Negative shell-correction energies are in blue and mean higher stability; positive shell-correction energies are in red and indicate destabilisation. The decay chain almost passes through the minimum of shell-correction energy located at proton numbers 108–110 and neutron number 162. The shell-correction energies are theoretical data and were calculated by Peter Möller. The upper panel shows the measured half-lives of the nuclei of the decay chain. There is a striking jump of half-lives by five orders of magnitude between the α-decays of ^{273}110 and ^{269}Hs (Z = 108).

heavy elements, so could we also trust its prediction of the spherical super-heavies? Unfortunately we could not. Thirty years had elapsed since the early years of calculations. Computers had become much, much more powerful, and more sophisticated methods were applied. Yet, the outcome remained confusing. Shells were predicted for the protons at Z = 114, 120 and/or 126, and for the neutrons at N = 172 and/or 184. The values depended on the type of model used but all of them were able to describe properties of lighter elements quite well; no single one of them was the best. We were left with vague ideas about where the superheavies could be found. The 'lake' in Figure 12.5, which starts to become visible on the 'horizon' on the right, may already belong to that region if the proton shell is at 114.

How could we get there? That was the third topic which remained unanswered, although we continued to make progress. We found that the predicted extra-push energy was not an absolute barrier for the synthesis of heavy elements and that heavy ions still fused at low excitation energy. The strong binding of the double magic ^{208}Pb as a target nucleus provides low excitation energies. Optimum beam energies are those which ensure that the nuclei just touch and then come to rest; fusion then starts spontaneously due to a rearrangement of the nucleons of the projectile around the lead core which proceeds in well defined quantum mechanical order. Friction, which would hinder the fusion process, is reduced to a minimum.

What did the production yield tell us about the possibility of producing superheavies (Figure 12.6)? The general trend decreases with increasing

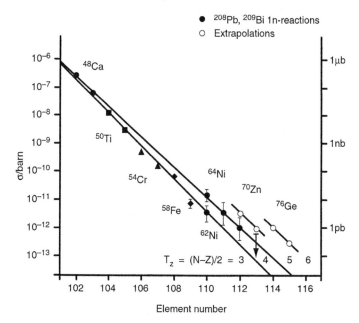

Figure 12.6 Plot of cross-section data for cold fusion reactions. The lines are labelled with the isospin number (T_z) of the projectiles, which is half the difference between neutron and proton number.

element number. The measured cross-sections start at a high value for the synthesis of element 102 with a ^{48}Ca beam and ends at element 112 at a value more than five orders of magnitude (100,000 times) less. With our apparatus, this meant one atom produced per week. Extrapolating the curve to element 114 loses two orders of magnitude, resulting in a production rate of one atom in 100 weeks (two years).

We had measured an increase in production yield in only one case, that of element 110 when we changed the beam from ^{62}Ni to ^{64}Ni. It seemed that more neutrons are useful for improving production but was it really the neutrons that helped, or was it the higher binding energy of the compound nucleus as N = 162 was approached? If the latter, we could hope that strong shell effects of the spherical superheavies would also reverse the trend of decreasing yield. Our most optimistic estimate for the production of element 114 using a ^{76}Ge beam was a value similar to the production of element 112. So there was a good deal of uncertainty as we again started to try to make progress with the superheavy nuclei. There was yet another matter: we had to decide how to name the new elements.

13

Names mean more than numbers

There is a Latin proverb *nomen est omen*: a name is a sign. Everybody and everything has a name: ourselves, our family and friends, animals and plants, things we see (cars, clouds and mountains), things we conceive (countries and philosophies), things we do (football or the Battle of Borodino) and things we 'remember' (dinosaurs and Neanderthals); stars still bear the names given to them in antiquity. Names express affiliations to a group, the family, the spirit of a time and names are taken from those whom we admire.

We need names so we know what we are talking about; they are so much more evocative and easier to remember. Suppose instead of *Mercury, Venus, Earth, Mars* and the rest we spoke of the planets just as no. 1, no, 2, no. 3 (that's us!), no. 4 . . . Life would be like telephone numbers.

Like everything else, names are given to materials. In ancient times the ones of importance were called water, ice, salt, iron, gold, etc. Some of them (iron, copper, silver and gold) turned out to be elements; others were actually compounds containing more than one element so new and more precise terminology was invented: thus, common salt, for instance is composed of sodium and chlorine and called *sodium chloride*.

As time went on, more and more elements were found in nature. After Mendeleev proposed the Periodic System of the Elements based on the measured properties of the 63 elements known at the time, chemists could search specifically for the gaps which showed up in the table. The last of them, *rhenium* (element 75) was discovered in 1925 and was named by the discoverers, the German chemists Ida Eva Noddack-Tacke, Walter Karl Friedrich Noddack and Otto Carl Berg working in Berlin, from the Latin *Rhenus*, the river Rhine.

The first synthetic transuranium elements were named after the planets: *neptunium* and *plutonium* (which follow the natural uranium) after Neptune and Pluto (beyond Uranus) – we mentioned that in Chapter 3. The new elements up to 101 were synthesised ('discovered' is not quite the word for what happened) in the US quite straightforwardly by Ghiorso, Seaborg and their colleagues: as we have seen, they were called *americium, curium, berkelium, californium, einsteinium, fermium* and *mendelevium* after famous chemists and physicists or famous locations where this sort of work had been done. But then things got more complicated, mainly around the synthesis of elements 104 and 105 and mainly because of insufficiently clear experimental results.

These complexities blocked the naming of the elements 107 to 109, all three of them without question synthesised at GSI in Darmstadt. The reason was that we at Darmstadt wanted to get general agreement, including that of our colleagues at Berkeley and Dubna, for naming the whole series of elements from 102 to 109. We thought it could be helpful to consider also names from those elements which had been named twice previously, like *rutherfordium* and *kurchatovium* for element 104, or *hahnium* and *niels-bohrium* for 105.

In 1985, IUPAP and IUPAC decided to establish a 'Transfermium Working Group' (TWG) to consider questions of priority in the discovery of elements above 100. An extensive report of their work was published in 1992 which officially confirmed our priority for the discovery of elements 107 to 109. We proposed the names in September, 1992, as part of a ceremony at GSI attended by Mike Nitschke from Berkeley and Yuri Oganessian from Dubna; it was almost ten years after our discovery of element 109. With support both from Berkeley and Dubna, we proposed *nielsbohrium* for 107, *hassium* for 108 and *meitnerium* for 109.

But there was still no agreement about elements below 107. One objection of IUPAC was that an element should not be named after a living person. This referred to element 106 for which the discoverers had proposed *seaborgium*. The name was announced in March, 1994 at a meeting of the American Chemical Society in San Diego by Kenneth Hulet (a retired chemist from Lawrence Livermore National Laboratory and one of the co-discoverers of seaborgium) in his acceptance speech for the ACS Award for Nuclear Chemistry honouring his lifetime achievements in the field. Seaborg had passed away on February 25, 1999, at the age of 85.

IUPAC was obviously getting tired of the arguments and, in 1994, proposed its own list, originating partly from the names proposed by the discoverers and partly from those of older propositions such as dubnium and jolotium for elements 104 and 105, respectively. They meant well but the discoverers said, 'No!' . All three laboratories protested. So much for trying to pour oil on troubled transuranic waters. Finally, in 1997, there was a general assembly of IUPAC at Geneva. The whole matter was reconsidered and an exception was made for seaborgium. Table 1.1 gives the names and abbreviations now officially in use.

Elements 110 to 112 were as yet unnamed and proposals were made to IUPAC and IUPAP. They had been synthesised within a short space of time between the end of 1994 and spring of 1996. We had not discussed names before the experiments and, during and after the irradiations, we were too busy to think about them. As a consequence, we did not, as had earlier been the custom, offer names in the reports of their discovery but left it for later. However, as soon as the discovery of a new element was made public, we received many suggestions, mostly serious, which we kept in an special folder for later review. Some others, though well-meant, were out of the question – or simply funny as, e.g., 'policium' for element '110', suggested by an American school class.

On October 15, 1997, I sent the following e-mail to my colleagues:

The names of the elements up to 109 were approved by IUPAC. We now have an opportunity to think freely about the naming of the new elements.

Since 1988, we have been working hard to make progress towards unknown elements. Our results are widely accepted so we felt we have the right and the duty to name the new elements 110, 111 and 112.

I propose that, during our next beam time in November/December 1997, we take the opportunity of most of us being here at GSI to organise a discussion on this subject.

I especially want to ask all of you to contribute suggestions so that we can prepare a list of candidate names and end up with the best choice.

With many greetings,

Sigurd

'Names-finding day' came in December 1997. From November 25 to December 18 we used a ^{50}Ti beam for various irradiations and the whole group was together again for the experiment. Two-thirds of the work had already been done so we arranged a meeting for Wednesday, December 10, 1997, at 2:00 p.m.

The GSI people were Peter Armbruster, Helmut Folger, Fritz Heßberger, Gottfried Münzenberg, Hans-Joachim Schött and myself. Anton Lavrentev, Andre Popeko, Sasha Yeremin were from Dubna, Stefan Saro from Bratislava and Matti Leino from Jyväskylä.

After opening the session, I had the pleasure to announce Stefan's birthday. That ensured that the mood of the occasion was light-hearted as we knew that, following the discussion, we would have an entertaining evening.

First of all we reviewed all the important dates in the discovery of the elements. We were aware that both Berkeley and Dubna claimed some evidence of having observed an isotope of element 110. Their actual physical evidence will come later; for the moment let us look just at the dates. The Berkeley experiment was done in August/September 1991, shortly before the SuperHILAC was shut down. However, the possible decay chain was actually found later during a further analysis of the data. Our first intimation was the draft of a paper sent to us on February 25, 1994, by fax from Al Ghiorso. He made an official announcement at a conference at Taormina early in June and his paper was published in February 1995 in the *Conference Proceedings in Nuclear Physics*.

In articles dealing with the discovery of a new element, it is argued that a claim for priority must be made in a publication submitted *to a refereed journal* (see Box 13.1). This indeed happened with the Berkeley results; their paper was received by *Physical Review C* on November 22nd, 1994, appearing as 'Evidence for the possible synthesis of element 110 produced by the $^{59}Co + ^{209}Bi$ reaction' in Volume 51 in May 1995. The important date for claiming priority is when the article is received by the editor. (Remember how in Chapter 12 we sweated writing our paper?)

During our naming meeting, we were indeed somewhat surprised when we reviewed the data. Our paper on element 110 had been received by the editor of *Zeitschrift für Physik* on November 11th, 1994; the editor of *Physical Review C* received the Dubna paper on January 19th, 1996. The Dubna paper was entitled 'α decay of $^{273}110$: Shell closure at $N = 162$'. From the title one already has the impression that the authors were not claiming the discovery of a new *element*, rather the discovery of a new *isotope*, $^{273}110$. For parallel reasons we chose as the title of our element 110 paper 'Production and decay of $^{269}110$'. This left the door open for establishing a discovery profile together with colleagues in Berkeley.

At Dubna, the search for $^{273}110$ (September 10th to December 30th, 1994) actually started before our run. We first heard about it via a phone call from Yuri Oganessian on New Year's Day, 1995. Shortly thereafter, in mid-January, Yuri Lazarev presented the results at our annual conference at Hirschegg in Germany.

We came to the conclusion that had been no unambiguous identification announced or published on elements 111 and 112 before our own papers, 'The new element 111' and 'The new element 112', received by the editors of *Zeitschrift für Physik* on December 21st, 1994, and February 21st, 1996, respectively. The titles of all three of our papers were simple and clear, their contents convincing.

After these formalities I went through the names proposed by various

---

Box 13.1 Refereed and unrefereed publications

Scientists have a number of legitimate routes to the publication of their findings:

- announce them at a conference and have them printed and circulated in the proceedings. As a rule, whatever written account the scientist submits will be published without change or comment;
- send them to a journal which uses expert referees to scrutinise every manuscript submitted to ensure that it makes sense, holds together, that the experimental procedures could indeed have generated the results reported and that those results justify the conclusions;
- send them to a journal (and there are some) which does not undertake detailed review.

Other possibilities are informal distribution of the data among workers in the field (but that does not establish priority) or go straight to the newspapers and broadcasting. Informing the public about new results is an essential task and considered as highly important by the scientists; however, in the case of usually complicated results a detailed presentation in a scientific journal is necessary. Earlier on (in Chapter 12) I showed how important it was for us to submit our scientific paper first and only then to think about how best to tell the public. A modern way of spreading news is to put the information on a 'web page'. But as everybody should know, not everything people present on their home pages is accurate.

---

people from other institutions and Andre wrote them on the blackboard. Without nicknames we had about ten proposals. Five reached us by e-mail from our young colleague Christelle Stodel writing from GANIL in Caen; she would prefer names of people who were still living.

I then asked colleagues in order round the table to give their views, after which I voiced my favourites. We ended up with a list of 30 names, mainly physicists of course, but not all men. There followed the great philosophers and natural scientists of the past. Some of the names represented towns, countries or important subjects thought of enduring value and still relevant for our own time. It is best to keep this list under wraps because some people will be upset when they find their favourite name is missing.

The discussion was very positive and I felt that during the afternoon we would be able to agree names for the three new elements. That would be another record: discussions of this sort are usually less harmonious, but this one revealed the good will within our group (Figure 13.1).

A critical moment came during the coffee break when Stefan brought in the wine. I feared that, if we drank it then, the discussion would go round in endless circles and never reach a conclusion. I asked him to keep the wine until the evening. Everybody agreed; they were beginning to feel that we were getting somewhere.

Still, it was difficult to select three names out of 30. At first, we crossed out names which might be kept for new elements in the future, leaving just a few in contention. Now the arguments raged fast and furiously. Painfully, one at a time, names were ringed with red chalk until, in the end, we had found the three we needed.

Then it was time to celebrate Stefan's birthday. In a more relaxed mood, we spoke about how to proceed further. Should we publish the names, and when? Was there a good opportunity in the near future, an anniversary or something? Did we have to ask anybody, relatives or institutions, before we could use a person's name for an element? Is there any sort of copyright on somebody's name?

Figure 13.1 Photo taken on December 10, 1997, at GSI at the end of our meeting to agree names for elements 110 to 112. From the left: Helmut Folger, Matti Leino, Alexander Yeremin, Fritz Heßberger, Peter Armbruster, Gottfried Münzenberg, the author, Stefan Saro, Andre Popeko and Hans-Joachim Schött.

No decision was made that evening about how to proceed further but a few weeks later we wrote to the representative of IUPAC, telling them our proposals for the names of elements 110, 111, and 112 and asking for them to be treated as confidential until they were officially announced; this was to avoid two or more names for element 110 suddenly circulating.

We had naturally exchanged views about the naming of element 110 with Al Ghiorso and Glenn Seaborg at Berkeley laboratory soon after our result on $^{269}110$ but some of the Berkeley people still wanted to use hahnium for element 105, rather than dubnium as accepted by IUPAC. We were unable to agree with Berkeley about a name for element 110, so IUPAC and IUPAP set up a small group of independent experts, the joint working party (JWP), to adjudicate the validity and relative value of the priority claims.

The report was published in 2000. We were really proud when we read:

> The redundancy of the consecutive alpha energies and delay times in the second through fourth chains measured is very reassuring. Even more so is the observation of fourth and fifth alpha particle energies and delay times in the last two chains observed that are in very good agreement with the known properties of descendants ^{257}Rf and ^{253}No.
>
> JWP ASSESSMENT: Element 110 has been discovered by this collaboration.

And the conclusion on the second isotope $^{271}110$ read: 'This study was very influential in the thinking of the joint working party. The high quality of the data and the internal redundancy in the nine chains served as a benchmark against which similar studies for this element and other elements were compared during the JWP deliberations.'

JWP kept their critical manner when they examined the studies of the heavier elements 111 and 112. Concerning 111 they assessed: 'The results of this study [our experiment] are definitely of high quality but there is insufficient internal redundancy to warrant certitude at this stage. Confirmation by further results is needed to assign priority of discovery to this collaboration.' And in the case of 112: 'The results of this study are of characteristically high quality but there is insufficient internal redundancy to warrant conviction at this stage. Confirmation by further results is needed to assign priority of discovery to this collaboration.'

We agreed to the conclusions, although it meant more hard work for us. Not only did we want to be convinced from the results, it was our aim to convince even the most critical colleague. Physics is not a question of believe or not believe, physics is a matter of facts. But this does not exclude that during the process of exploration stages are passed which are uncertain and full of contingencies. To judge the correctness of results is often the most difficult part in research work. We would soon realise the truth of this statement from our own work, as described in the following chapter.

We repeated the 111 experiment in October 2000. During an irradiation time of 13 days, we measured three more decay chains in addition to the three existing ones. They fully confirmed the previous results. The quality of the data was even improved, because all three chains were measured in a period when the projectile beam had been switched off immediately after the detector implantation of a nucleus, which was identified by an electronic circuit as a candidate for a 111 fusion product.

The synthesis of element 112 was repeated in May 2000. The experiment lasted 19 days and one more decay chain was measured, which agreed with chain number two measured in the 1996 experiment down to the decay of ^{261}Rf. There, however, instead of the previously measured α decay, now spontaneous fission was observed. This was a new result and not reported from older experimental work on that nucleus, although theoretical predictions are not in contradiction to a spontaneous fission branch of ^{261}Rf . Unexpectedly we got support for our result from our chemist colleagues, who had studied the decay chain of ^{269}Hs in May 2001. We already mentioned this experiment in Chapter 10. The nucleus ^{269}Hs, which decays further down to ^{261}Rf, is the granddaughter of ^{277}112. The chemist's result was that indeed ^{261}Rf has a spontaneous fission branch and, furthermore, that the nucleus ^{269}Hs, which we had identified as a member of a decay chain, also belongs chemically to element 108.

After the successful confirmation experiments we are confident that the JWP will agree to our results and eventually will approach us to suggest final names.

Although the naming of an element is not really a matter of high priority, the people involved do feel very strongly about them. Partly, I suppose, it is because everyone feels qualified to give an opinion and partly because they become very proprietary and emotionally involved. It is, after all, the very centre of their lives for the people doing it. Is being first that important? I think not – but others may think differently, particularly if prestigious honours may follow.

14

Attack on the domain of superheavies: elements 114, 116 and 118

The radiative capture channel and search for element 116

One of the unresolved questions giving us most headaches was the possibility of synthesising superheavy nuclei by a complete fusion of projectile and target nucleus without the emission of a neutron. The excitation energy would be emitted only as γ-rays, hence the name radiative capture (or '0n' meaning zero neutron) channel. We knew that this reaction channel was possible for heavy ions because some years earlier Karl-Heinz and co-workers had observed ^{180}Hg synthesis by fusion of two ^{90}Zr nuclei. In that reaction, the 0n channel had a cross-section only slightly smaller than the 1n channel. In the region of heaviest elements the step from the 1n channel to the 0n channel would be a major change in the properties of fusion reaction, but in which direction – larger or smaller cross-section?

We had found that the production yield for heavy elements is shifted towards smaller excitation energy of the compound nucleus with increasing element number. For element 112 we did not have a definite answer because only one beam energy was used from which we could not determine a maximum. But in the case of element 110 we had measured three energies and the highest yield was observed at an excitation energy only slightly higher than the neutron binding energy. We calculated that, for the synthesis of element 116, the excitation energy would be below the neutron binding energy, for element 118 it would be zero and for element 126 very negative. However, negative excitation energies are energetically forbidden and in that case it would be necessary to compensate for the negative energy by a surplus of beam energy (Figure 14.1).

If the excitation energy were to drop below the neutron binding energy, only γ-radiation would have to compete with fission. However, the probability of fission would be much reduced because of the low excitation energy of the compound system, giving the 0n channel an unusually high cross-section. Why not search for this channel for heavy elements?

We selected a reaction for which we expected an excitation energy at the fusion barrier smaller than the neutron binding energy:

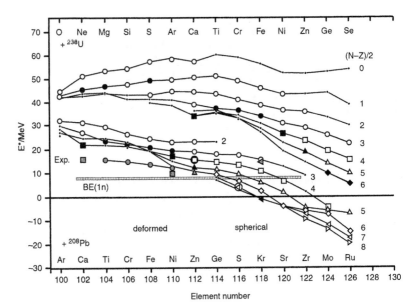

Figure 14.1 Diagrams of excitation energies for the synthesis of superheavy elements in fusion reactions using targets of ^{208}Pb (lower cluster of curves) and ^{238}U (upper cluster of curves). The projectiles are selected so that compound nuclei are produced of elements from $Z = 100$ to 126. The element name of the projectile is given for the reactions with ^{208}Pb target in the lower part, for the reactions with ^{238}U target in the upper part. For each projectile element various isotopes are used. Projectile nuclei with the same difference of neutron minus proton number divided by two, $(N - Z)/2$, are connected by lines; these differ by just one α-particle. High values of $(N - Z)/2$ mean neutron-rich nuclei and low values neutron-deficient nuclei. In general, the coldest compound nuclei (low excitation energy) are produced with the most neutron-rich nuclei. However, if the projectiles carry too many or too few neutrons, they became radioactive. Those radioactive projectiles are marked by the small dots. The diagrams impressively reveal the difference between cold (using ^{208}Pb target) and hot (using ^{238}U target) fusion reactions. The maximum yield for production of elements from 102 to 110 was measured at excitation energies even below the calculated values. The experimental data are marked by the hatched symbols.

An important transition occurs between element 114 and 116, where the neutron-binding energy is crossed; the neutron-binding energy is shown in the plot by the straight line BE(1n) at $E^* \approx 7.5$ MeV. The result on element 118 in Berkeley was obtained using the reaction ^{86}Kr + ^{208}Pb → ^{293}118 + 1n at an excitation energy of 13.3 MeV; that data point is also marked. The value $(N - Z)/2$ equals 7 for ^{86}Kr. At this point the calculated excitation energy for the reaction is already slightly negative. The secret of the relatively high cross-section for the synthesis of

^{82}Se $+ \ ^{208}$Pb $\rightarrow \ ^{290}$116*; that ^{290}116 (N = 174) nucleus would also be a close approach using cold fusion to the double magic nucleus Z = 114, N = 184. The location of the nucleus ^{290}116 can be identified in the Figure 10.1 which shows an extended version of the chart of nuclei up to element 120. We are now entering the domain of the superheavies.

Six beam energies were used for 33 days in November/December, 1995, but nothing turned up. What we had thought a great idea was dead, at least temporarily, but we retained the thought that this channel would be an option for future experiments aiming at the synthesis of SHEs.

Search for element 113

The successful synthesis of element Z = 113 would be another step towards the discovery of spherical superheavy nuclei. However, the search was complicated by the fact that no excitation function for the production of odd elements from dubnium (Z = 105) to element 111 was well enough known to permit an estimate of the optimum beam energy. We therefore investigated the two reactions ^{50}Ti $+ \ ^{209}$Bi $\rightarrow \ ^{258}$105 $+ 1$n and ^{58}Fe $+ \ ^{209}$Bi $\rightarrow \ ^{266}$109 $+ 1$n before the main experiment. The most important outcome was that the position and width of the excitation function for dubnium and for meitnerium were the same as for the next lighter even element. Knowing the beam energy for production of element 112 would mean also knowing the beam energy for production of element 113.

Using that beam energy, we started the irradiation at SHIP on March 5, 1998. In 25 days, we collected a beam dose of 4.5×10^{18} projectiles and another 3×10^{18} projectiles during 21 days in April. In total this was twice as much beam dose as we had had for the synthesis of element 112, but no 113 showed up. The value of our cross-section limit is shown together with other values in Figure 12.6.

---

element 118 was explained by the decrease of calculated excitation energy with increasing element number and as also indicated in the plot by the transition of the shape of the superheavy nuclei from deformed to spherical at about element 112. Two aspects are important in the case of hot fusion: 1, the local minimum excitation energy using ^{48}Ca beam, (N − Z)/2 = 4, for the synthesis of element 112; 2, the decrease of the excitation energies to values near the 1 neutron binding energy for the synthesis of element from 122 to 126 using beams of ^{70}Zn to ^{82}Se. The study of these reactions was considered at GSI and no doubt elsewhere as well just after the Berkeley result on element 118.

The new Dubna strategy

Cold fusion experiments using targets of lead and bismuth were first performed with the U-400 cyclotron in Dubna. Significant results, including the
investigation of element 109, had earlier been obtained there (Chapter 6) but
later the Dubna group changed their strategy and resumed the use of hot
fusion using actinide targets. We at GSI were well informed about the Dubna
plans; everything of collective interest was very openly and freely discussed
as we worked together or met at conferences. The reason for the Dubna
changes came about

- because of the low cross-sections (due to increasing Coulombic repulsion
  – the extra-push prediction), expected for the production of SHEs by cold
  fusion;
- from the fact that only actinide targets allowed the production of longer-
  lived, more neutron-rich nuclei closer to the region of the spherical SHEs;
  and
- because lighter beams were easier to produce from the cyclotron with
  high intensity.

The lowest excitation energies of compound nuclei resulting from
fusion with actinide targets are obtained with beams of ^{48}Ca; Figure 14.1
clearly shows the local minimum for the excitation energy in reaction with
^{48}Ca and ^{238}U. Because of the high neutron number of ^{48}Ca, the most
neutron-rich nuclei are accessible. The development of an intense ^{48}Ca beam
with a low consumption of material in the ion source was the aim of the
Dubna work over about two years in 1996–8.

The U-400 cyclotron was modified to allow for continuous operation
and the beam intensity increased 2–3-fold. These gave an average intensity at
the target of 2×10^{12} ^{48}Ca ions/second at an extremely low consumption rate
of only 0.3 milligrams per hour; with the current world market price for ^{48}Ca,
this amounted to only $30 per hour. Upgrading the cyclotron was certainly a
good investment.

The U-400 cyclotron experiments were performed at two different
recoil separators built during the 1980s. One is VASSILISSA (see Chapter 11),
an electromagnetic separator similar to the one at SHIP. The other is a gas-
filled separator named 'GNS' which stands for '*G*aso *N*apolnenny *S*eparator',
Russian for gas-filled separator (Figure 9.1). Both separators had also been
upgraded to improve efficiency and background suppression. The detectors
were almost an identical copy of our system at SHIP (shown in Figure 11.6).
Extensive use was made of fast and relatively cheap PCs for data acquisition
and analysis. Nobody commented on the old days when fast VAX computers
were embargoed and could not be exported to the Soviet Union.

Both separators, constructed and maintained by two distinct, non-overlapping groups, were mounted in an experimental hall close to the cyclotron at parallel beam lines. The group organisation may have a historical base and certainly reflects the behaviour of scientists: competition among them is a powerful stimulus for the development of science to be set against a readiness to help colleagues. The balance between these tendencies is determined by its group dynamic. It all seems to me to be very complicated, almost as difficult to describe and predict as the weather!

The fact is that the VASSILISSA group collaborates with the SHIP group while the GNS group works closely with scientists from the Lawrence Livermore National Laboratory in California. One might think that this is a tactically brilliant strategy on the part of the Dubna laboratory leaders but I suspect it is simply the result of who knew whom and how the groups worked internally.

Experiments at the gas-filled separator GNS

The properties of the gas-filled separator are such that fusion products from very asymmetric reactions (i.e. using relatively heavy actinide targets and relatively light projectiles) are separated with slightly higher transmission than in evacuated separators. At the beginning of the 1990s, Yuri Lazarev and his colleagues had therefore started a programme to look at hot fusion reaction using actinide targets; that was when cold fusion was predicted to have no future. Possible actinide targets are the natural isotopes of thorium and uranium (^{238}Th, $Z = 90$, ^{238}U, $Z = 92$) and lighter isotopes of uranium as well as the synthetic isotopes of plutonium (^{242}Pu and ^{244}Pu, $Z = 94$), americium (^{243}Am, $Z = 95$), curium (^{248}Cm, $Z = 96$) and californium (^{249}Cf, $Z = 98$).

The first new result using the new GNS set-up was the identification of the hitherto unknown isotopes ^{265}Sg and ^{266}Sg. The reaction was ^{22}Ne + ^{248}Cm → ^{270}106*. The compound nucleus was hot (highly excited) and five and four neutrons, respectively, had to be evaporated for the production of the new isotopes. Only four atoms of ^{265}Sg and six of ^{266}Sg were observed. Both nuclei decayed by α-emission; the measured α-energy was low but no longer surprising as theory predicted low α-energies for nuclei close to $N = 162$. The experiment confirmed that.

Low α-energy means long half-life and that was what the nuclear chemists were waiting for. Unfortunately half-lives could not be measured in the Dubna experiment because the implantation of the seaborgium nuclei into the detector was at such low energy that no signals could be obtained indicating the birth of the new isotopes. However, from the α-energy one could make an estimate of the half-life using well known energy half-life relations; that gave an estimate of several tens of seconds, creating a good deal of excitement among the nuclear chemists because that would be ideal for investigating the chemical behaviour of element 106 and seaborgium chemistry, which was indeed later studied at the GSI. Concerning the nuclear aspects, the Dubna data for the half-lives were confirmed at about 10 seconds. It was shown chemically that seaborgium behaves typical as a group VIA element like molybdenum and tungsten.

Back to the Dubna GNS. The next step after synthesising element 106 by using hot fusion was to attempt to make element 108. Lazarev and co-workers successfully did that, using $^{34}S + ^{238}U \rightarrow ^{272}108^*$. The new isotope $^{267}108$ was identified by parent/daughter correlation, a total of three decay chains being measured. In November 1994 we confirmed the decay properties of $^{267}108$ at SHIP, measuring this nucleus as daughter of the α-decay of $^{271}110$.

In 1994, an unforeseen race started between Dubna and Darmstadt over the new element 110. The contenders were hot fusion at Dubna, starting in September, and cold fusion at Darmstadt, beginning in October (see Chapter 12). The history of discoveries is full of examples where nothing has happened perhaps for hundreds of years and then, suddenly, the time becomes ripe and the project is attacked simultaneously by several people. The tragic race to the South Pole between Amundsen and Scott was just such a case; luckily our race was not quite so dangerous – or so cold – although to have lost it would have been depressing, to say the least.

In Chapter 12, I have already described how the race was won in Darmstadt, but Dubna was successful too. They had needed time for the analysis of their data. On New Year's Day, 1995, Oganessian tele-phoned and informed Gottfried and me that one α-decay chain had been observed which was assigned to the nucleus $^{273}110$ having been pro-duced in the reaction $^{34}S + ^{244}Pu \rightarrow ^{273}110 + 5n$; the production cross-section was about 20 times less than in the cold fusion in Darmstadt. The Dubna isotope had two more neutrons than our $^{271}110$ so we could not directly compare the decay data. Later, from our data on the decay of $^{277}112$ which produces $^{273}110$ as a daughter nucleus, we did not get 100% agreement but the difference might be explained by the complicated decay pattern of this nucleus.

Although our hunting of elements was competitive, the relations between the people in Dubna and Darmstadt was that between colleagues working in the same field. There were no secrets, results were exchanged and our plans were discussed just as freely as ever. For example, the GSI beam time schedule is accessible to everybody on our web site and I also officially informed Lazarev of our finding as soon as the paper was submit-ted. He answered by e-mail as follows:

Date: November 16, 1994

Dear Sigurd,

Thank you very much for the 110 news and please receive my sincere congratulations. For me personally, the most impressive thing in the whole of your work was your ability to produce and submit the paper in so short a period of time, just five days. Let me joke that for A. Ghiorso and co-workers it took about one thousand days. So, you evidently have a gain of two orders of magnitude.

Speaking seriously, I wish you good luck with new 110 events, as well as many forthcoming Short Notes to Zeitschrift fuer Physik A.

By now, we have collected a total beam dose of 1.2*10**19 for the 244Pu+34S reaction. We are starting the data analysis. The second 244Pu target wheel proved to be much more stable, so we continue our bombardment at a good pace. I will inform you of the analyses outcomes as soon as we have firm conclusions.

Best regards.

Yours,

Yuri Lazarev.

In January, 1995, I sent my own congratulations to Yuri (I could not do so earlier because the computer in Dubna was shut down over the Orthodox Christmas):

Dear Yuri,

Congratulations on your successful 110 experiment.

I got the news from Yuri Oganessian on our New Year's day, Jan. 1st. I think, for all of us – the low energy heavy element fusion people – it was a great year.

There is certainly a good chance that you will find more in the data.

I wish you and your colleagues good luck, and a happy New Year 1995.

Best regards,

Yours,

Sigurd

Sadly, and completely unexpectedly, Yuri Lazarev died early in 1996; the group in Dubna had lost a leader and the heavy element community a competent scientist and colleague.

For insiders it was clear that, after the 110 experiment, the people in Dubna would like to go on to 112 and 114. An attractive projectile-target combination would be $^{48}Ca + ^{238}U$ for element 112 and $^{48}Ca + ^{244}Pu$ for element 114. This aim made the development of an intensive ^{48}Ca beam essential; it took until the autumn of 1998 to get it ready.

News from Dubna: element 114 discovered

The work at the cyclotron U-400 was finished in 1998. In November–December, an experiment was performed at the GNS aiming at the synthesis of element 114, a target of ^{244}Pu being irradiated with a ^{48}Ca beam for 34 days.

On December 30, 1998, I received a private message from Dubna, the day before it was announced officially: on November 29th the Flerov labora-

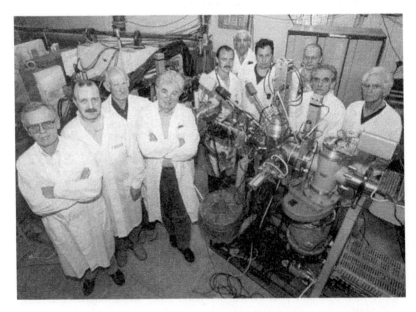

Figure 14.2 Research team at the Dubna gas-filled separator GNS. The separator is in the top-left of the photograph and the protruding arms of the target apparatus are visible toward the right. The group includes the leading researchers Utyonkov (second from left) and Oganessian (fourth from left). At the time the collaborators from Livermore had already left the Flerov Institute. The photograph is taken from *Scientific American*, January 2000.

tory had observed one event chain assigned to the decay of ^{289}114. That decay chain and the compound nucleus ^{292}114 at $N = 178$ are shown in Figure 10.1 and the successful GNS team in Figure 14.2. Our Dubna colleagues sent us detailed information and I immediately started to reconstruct the decay chain and to compare it with theoretical predictions. The chain comprised an implanted nucleus followed by three α-decays and subsequent fission. It was actually the fission event which enabled the Dubna people to find the chain quickly. No doubt it was Oganessian who took into account the calculations suggesting that the daughter nuclei would undergo fission after a few α-decays. The Dubna strategy for analysing their mass of experimental data was, firstly, to select the few fission events which could be easily found because of their high energy, and then search the data already on file. In this way, the three α-decays and the implanted parent nucleus were easily identified – just as we had identified ^{266}109 at SHIP in 1982. In the middle of January, 1999, after the Orthodox Christmas and New Year, we were officially informed by e-mail from Oganessian. The paper entitled 'Synthesis of superheavy nuclei in the ^{48}Ca + ^{244}Pu reaction' was submitted by the Dubna–Livermore collaboration to the journal *Physical Review Letters* on March 8th, 1999.

The properties of the chain were in themselves consistent, meaning that the α-energies corresponded to the lifetimes; that had also been checked at Dubna. The decay pattern of the chain followed the theoretical predictions but I was still not entirely convinced that element 114 had really been discovered. Several of my colleagues thought similarly, so we did not experience the euphoria which usually comes with a great discovery.

I worried about the following points:

1. the lifetimes were very long, 30 seconds, 15, 1.6 and 16 minutes for ^{289}114, ^{285}112, ^{281}110 and ^{277}108, respectively. Might the signals not be due to adventitious background signals?
2. the α energies were low at about 9 MeV. Could they have originated from reactions with target impurities?
3. the cross-section was relatively high, twice as high as that for element 110 produced earlier in Dubna by hot fusion and identical with that for the synthesis of element 112 using cold fusion. How could there be a higher yield for element 114 than for 110?
4. and finally, the new chain was far from any known nucleus. It was located in a completely white area in the chart of nuclei, shifted eight neutrons to the right from our element 112 decay chain. Based only on the calculated excitation energy, was not the assignment to the 3 neutron evaporation channel very vague?

It was. We discussed all these questions with our colleagues in Dubna who, of course, had also thought carefully about these possibilities. Although a reasonable explanation could be found for each of the questions, complete confidence and certainty were impossible. We were left with the clear need for further experiments, something not too easy when it takes 40 days of irradiation to observe one event.

Nevertheless, in 1999, several events quickly followed one another. In March, another lighter isotope of element 114 (^{287}114) was observed at VAS-SILISSA in irradiation of a ^{242}Pu target and, in May, news came of the synthesis of element 118 in Berkeley.

That spectacular event influenced the strategy at Dubna's GNS. They had to repeat the reaction ^{48}Ca + ^{244}Pu $\rightarrow$ ^{289}114 + 3n, first to make quite sure and second to claim priority for element 114 because that had also been observed in Berkeley as a granddaughter of 118 decay. The VASSILISSA result (^{287}114) was received by the scientific journal *Nature* on April 19th, 1999, also before the Berkeley publication. Although the Dubna people had now two element 114 irons in the fire, ^{289}114 and ^{287}114, both published prior to the Berkeley decay chain, confirmation was needed in order to get acceptance. Once more we see that fast publication is highly recommended in this business.

To continue with GNS by way of an aside: the experiment there ran for the 4.5 months from June to October, 1999, longer than our great SHIP run of 1994 which was 'only' 2.5 months. And the result? The first chain was not found again but two other almost identical chains were. The first was measured on June 25th, the second on October 28th. Both started with an implanted nucleus, followed by two α-decays, and finally a fission. This time the lifetimes were shorter, in the range of 1–20 seconds. These two chains were definitely different from the first; they were assigned to the 4 neutron evaporation channels, i.e. to the isotope ^{288}114. The location of the two chains in the chart of nuclei can be seen in Figure 10.1.

The production yield was comparable to that of ^{289}114 and the decay data had also been closely predicted theoretically. The short lifetimes made the accidental creation of the chains highly unlikely and the new result looked very convincing. A paper was sent for publication to *Physical Review Letters* at the end of December 1999.

But where was the 3n channel? The beam dose was twice as high in the second run as in the run at the end of 1998 but that is not necessarily a problem because statistical fluctuations are inevitable for such a small number of events. The beam energy was the same in both experiments so why was the 3n channel observed in one and the 4n in the other? Is not higher energy needed for the 4n channel? But it was not too difficult to find an explanation for this apparent discrepancy: there is an energy region just at the starting of the fusion process where both channels have similar production yield.

What remains as a possible misinterpretation is a slight uncertainty about whether only neutrons were evaporated. Perhaps a proton or an α particle had been involved making the residue not element 114 but 113 or 112; final proof will come only with more data. Moreover, the new decay chain is not connected to any known isotope but does have two neighbours, one from $^{289}114$ and the other from $^{287}114$. The discovery of the latter will be described in the next section.

Experiments with the electromagnetic separator VASSILISSA

The performance of VASSILISSA was tested in a series of experiments aimed at the identification of new neutron-deficient isotopes of uranium, neptunium and plutonium. Hot fusion reactions leading to the compound nuclei ^{258}No and ^{263}Db were studied in detail, resulting in new data and serving also as preliminary experiments for the more difficult studies of the heavier elements at much lower production yield. The value of such tests cannot be counted high enough.

Since everything seemed to be working well, attempts were begun to search for new isotopes of element 112 by irradiation of ^{238}U with ^{48}Ca ions. The irradiation started in March, 1998, and, over a period of 25 days, two fission events were measured with a mean half-life of 81 seconds and a cross-section five times larger than for the synthesis of $^{277}112$ at SHIP. The two events were tentatively assigned to the residue $^{283}112$ after 3 neutron evaporation. In a second run at a slightly higher beam energy, in which one would expect 4 neutron evaporation, no heavy nucleus decay events were observed.

Using the reaction $^{48}Ca + {}^{242}Pu \rightarrow {}^{290}114^{*}$, experiments at VASSILISSA were continued in March, 1999, following the 114 experiment at the GNS. It was expected that, after evaporation of three neutrons, the nuclide $^{287}114$ would be produced and would decay by α-emission into $^{283}112$ which had already been studied. Over a period of 21 days, a total of 7.5×10^{18} projectiles were collected with a beam energy setting to optimise the emission of three neutrons from the compound nucleus.

Four fission events were detected. Two could be assigned to known fission isomers, tentatively to the ^{241}Pu (half-life 24 microseconds) produced by neutron transfer. Fission isomers are quasi-stable, relatively long-lived and strongly deformed states in nuclei which decay by fission. They were discovered in 1962 by Sergei Mikhailovich Polikanov in Dubna. Fission isomers occur only in a limited region of nuclei from neptunium to berkelium and they can be produced in relatively large quantities when actinide targets are used. There is certainly a danger of misinterpretation when fission is assigned to heavy elements produced in fusion reactions. The two fission signals were registered 59 and 20 microseconds after implantation, in agreement with the half-life of the fission isomer of ^{241}Pu.

The other two fission signals were preceded by signals from α-particles and implanted nuclei. The signals were assigned to the nuclide ^{287}114. The mean values for the lifetimes of nuclei obtained from the two chains are given in Figure 10.1.

The four events, two from 112 and two from 114 produced by irradiating ^{238}U and ^{242}Pu, respectively, with ^{48}Ca are consistent. The fission lifetimes are within the limits given by statistical fluctuations. Fission was again measured after α-decay, when the target was changed from ^{238}U to ^{242}Pu. The low background rate in the focal plane of VASSILISSA makes production by chance coincidence unlikely. The excitation energy was ideal for the evaporation of three neutrons but did, however, also allow for the emission of an α-particle or a proton. Although the evaporation of charged particles is highly hindered, that possibility cannot be excluded completely.

VASSILISSA is at present being further upgraded. The last dipole magnet is being replaced by a magnet of higher bending power which will result in a mass resolution of about 1.5%. This additional information will considerably limit the range of possible masses for the implanted nuclei. It is especially important in cases where the decay chains do not end in a region of known nuclei, where no α's are emitted and the implanted evaporation residue already undergoes fission or where the lifetimes are extremely long (greater than about one hour), as expected for some superheavy nuclei. This is certainly a good investment into the future.

News from Berkeley: element 118 discovered

After the shutdown of the SuperHILAC linear accelerator in Berkeley, 1992, ideas for building a new efficient separator for fusion-reaction products at the 88-inch cyclotron were discussed. The first design by Albert Ghiorso of a *Large Angle Separator SYstem* (LASSY) was rejected. This was to be a gas-filled separator utilising superconducting magnet technology. The high cost and prolonged construction time led to a parallel effort to design the Berkeley Gas-filled Separator (BGS), one based on normal-conducting magnet technology which could be built much more economically and much faster. A design study had been made by Kenneth E. Gregorich, Ghiorso and co-workers in 1995–6. This was at the time when Ghiorso, in the analysis of his last run at the SuperHILAC, had found a rudimentary decay chain which he believed could be assigned to ^{267}110. He wanted to repeat that,

^{59}Co + ^{209}Bi → ^{268}110* irradiation, using a new and better separator at the 88-inch cyclotron.

The final version of the BGS was designed and managed with the help of Victor Ninov, who had left the SHIP group at the end of 1996. He took up a position in Berkeley which had become vacant following the death of J. Michael Nitschke, one of the most competent physicists at the Berkeley laboratory and co-discoverer of element 106; he appears in Figures 5.2 and 10.10. For Victor it was a chance to take part in the construction of a gas-filled separator which was actually an extension of his thesis work. Later he might hope to develop his own experimental programme, which would have been not so easy in GSI's already overloaded research programme. I was very sympathetic with his decision and we supported him in Berkeley with computer programs, targets and detectors.

The BGS consists of three magnets of a special design so that the transmission of reaction products is very high. The bending angle of 70° is unusually large but provides high background suppression for reactions with lead targets up to the heaviest beams (about that of krypton) which can be delivered by the cyclotron. A scheme of the BGS is shown in Figure 14.3.

The BGS experiments were ready by 1998. In the autumn of that year, Robert Smolanczuk was working on a theoretical study of the production mechanism of superheavy nuclei in cold fusion reactions. Robert was a member of a group of theoreticians headed by Adam Sobiczewski in

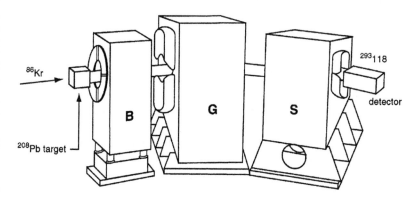

Figure 14.3 Scheme of the Berkeley Gas-filled Separator (BGS) used for separation of element 118 in the reaction ^{86}Kr + ^{208}Pb → ^{293}118 + 1n. The three ion optical elements are, from the left, a vertically focusing quadrupole magnet, followed by a strong horizontally focusing gradient dipole magnet and finally a flat field dipole magnet. The total deflection angle amounts to 70°. The figure was based on Ninov, Gregorich and McGrath (1998).

Warsaw, one of the first groups calculating the stability of superheavy elements. Robert was using his computer codes and experience gained in calculation of fission half-lives for the inverse process, the fusion of heavy ions. Using a relatively simple model, he reproduced the measured formation cross-sections of deformed heavy nuclei synthesised in cold fusion reactions with elements 102 to 112. The same model, applied for the calculation of cross-sections for the synthesis of spherical superheavy nuclei, resulted in an unusually high value for the reaction $^{86}Kr + ^{208}Pb \rightarrow ^{293}118 + 1n$. He found the yield for the production of element 118 to be 670 times higher than for element 112. He did those calculations during a six-month stay in the SHIP group at GSI so we knew very well what he was calculating but did not believe his results.

We were not the only ones to feel that Robert might be wrong; we had reason for thinking that he might have overestimated the yield. First, we had measured decreasing yields up to element 112. Why then should the yield be almost 1,000 times larger for element 118? Second, we had already tried – and failed – to produce element 116. Why should the synthesis of element 118 work any better? Third, calculations from other theoreticians predicted a cross-section for element 118 smaller than that for element 112 by a factor of about 200. Finally, our aim was to search for element 114 after systematic preparation by the measurement of the excitation function for element 112 from which we expected to obtain optimal beam energy. Each additional beam time would have cost time and would have delayed our programme. We wanted to proceed systematically in small steps, a successful approach until then. Too much beam time had already been wasted in earlier irradiations looking for superheavy elements. However, we were also aware that, if strong shell effects existed at $N = 184$ and $Z = 114$ to 126, they could influence the fusion process in an unforeseeable way. However, where exactly was the proton shell? It would have been too expensive and too time-consuming to check all possible combinations resulting in elements from 114 to 126.

For completeness, I want also to mention that after Smolanczuk publicised his result the people in Dubna were also thinking about testing the reaction $^{86}Kr + ^{208}Pb \rightarrow ^{294}118^*$ but they had just started the ^{48}Ca programme and also, in VASSILISSA, would have had difficulty in obtaining a voltage high enough for separation of the relatively heavy krypton beam, despite the difficulty to get an intensive beam from the cyclotron. At GNS the background would probably be too high, so the experiment was abandoned.

The Berkeley group thought differently. They had just finished their BGS when Smolanczuk came up with his prediction. The beam of the noble gas krypton could easily be produced from the ion source available at the cyclotron and they had beam time available, just five days from April 8th–12th, 1999. They, too, were not convinced by Smolanczuk's results but they might have had a vague hope. If his predicted cross-section were correct, they might expect that a few hundred element 118 atoms would emerge from the target during the five days. If not, then at least Smolanczuk would have been disproved by experiment.

The beam energy was selected so that one neutron would be evaporated from the compound nucleus. The lead targets were delivered by GSI and mounted on a wheel rotated at 400 rpm. The separator efficiency for this reaction was estimated as 75%. The detector was borrowed from the SHIP resources: a position-sensitive silicon strip detector was used as a central detector but no detector was mounted in the back hemisphere to measure escaping α particles. In front of the silicon detector, a parallel plate avalanche counter was used instead of secondary electron foil detectors. Behind the silicon detector, a second silicon strip 'punch through' detector was installed to reject particles passing through the primary detector. The dead time of the data acquisition system was relatively long at 120 microseconds. The whole system was not yet in its final form but it worked surprisingly well.

Two event chains, consisting of an implanted heavy nucleus and of subsequently emitted α-particles were measured during the five days of irradiation.

The experiment was repeated in Berkeley April 30th–May 5th, 1999. During the six days of that irradiation one more chain was observed. The cross-section resulting from the two parts of the experiment was not as large as Smolanczuk had predicted: indeed it was about 300 times less but nevertheless twice as large as the cross-section for the synthesis of element 112. The mean values calculated from the three chains are given in Figure 10.1. Not only was one 'baby born' (as Ghiorso liked to call the discovery of an element) but there were twins, elements 118 and 116. And if the Dubna results on element 114 would turn out to be wrong, even triplets.

We heard about all that later; at the time we knew only that the irradiation had been performed. I was at a conference in Dubna at the end of April and the information we had from Berkeley was that nothing had been seen with a cross-section of more than 5 picobarn (actually the cross-section from the first part of the experiment was 4 picobarn). All the Dubna and Darmstadt people could still sleep peacefully. If only they had known the truth . . .

I got the news on Friday, May 28th, 1999, by fax from colleagues at the Nuclear Chemistry Institute of the University in Mainz. They had received the message, also by fax, the day before, not directly from Berkeley but from somewhere else in the US. May 27th was also the date when the paper was received for publication by *Physical Review Letters*. I needed half an hour or more to read the paper carefully; it looked quite convincing and I informed our director accordingly. He was very happy.

We carefully examined the experimental data. They suffer slightly from the fact that no detectors were mounted in the back hemisphere, so seven of the 19 signals did not contain the full α-energy. Nevertheless, the signals were strong enough to provide time information and allow for a reasonable assignment to α-particles. It is very likely that the first α-decay of the third chain was lost because of a relatively long dead time. The α-energies of the decays assigned to $^{293}118$, $^{285}114$, $^{281}112$, and $^{277}110$ reveal some internal redundancy, each appearing twice within the limits set by the detector resolution. The lifetimes up to the fifth α-decay are short, on the order of one millisecond, which makes an origin by chance coincidences extremely unlikely. Electric disturbances are excluded by the authors (private communication). The assignment of the events to the decay of $^{293}118$ is most likely at the low excitation energy of the compound nucleus.

The reaction $^{86}Kr + ^{208}Pb \rightarrow ^{293}118 + 1n$ investigated at SHIP

As news of the Berkeley results spread, something interesting happened psychologically. Most people had in mind only the high value of the 'calculated' cross-section, not the 300-fold smaller 'experimental' value. The mean value from both parts of the Berkeley experiment was 2 picobarn, only twice that obtained in the rather difficult experiment for element 112. As a consequence, those people thought that making superheavy elements was 'easy' and preparations for doing so were even started in smaller laboratories. (It reminded me rather of the worldwide flurry of activity when 'cold fusion' was 'discovered' in 1989.) The heavier elements 120, 122 and also the odd elements from 119 upwards now seemed accessible and easy to make. Even at GSI we were not spared such speculation but, after few days, most of us once more recovered our sense of realism. The opinion gained ground that first we should try to reproduce the Berkeley result at SHIP.

We were quick to do so, starting our experiment as early as June 7th, 1999. We used the same reaction and beam energy as in Berkeley, which resulted in the same excitation energy of the compound nucleus. After nine days, no event. If we had seen just one chain, this would have meant a cross-section of 1.6 picobarn, below the Berkeley value. However, included statistical fluctuations, the limit was 2.8 picobarn, still larger than the Berkeley value. We had to go on.

The beam time schedule allowed for a second part of the experiment in July–August, 1999. This time we wanted to be as safe and sure as possible. For the purpose of controlling our set-up, we first repeated the reaction

^{58}Fe + ^{208}Pb → ^{265}108 + 1n, by then well known, measuring a total of 16 events during two days. The resulting cross-section was in agreement with the previously known maximum value. The test demonstrated that the UNILAC energy and the electromagnetic field settings of SHIP are reproducible and that the analysis programme was working properly. In a second step we compared the beam energies of the Berkeley cyclotron and the UNILAC. The result left us with some uncertainty but was nevertheless acceptable.

The main part of the second irradiation started on July 26th. We ran for 15.5 days, again without observing any chain similar to those seen in Berkeley. The cross-section limit resulting from both parts of the experiment was 1 picobarn.

We were therefore unable to confirm the Berkeley data on the synthesis of ^{293}118. However, our negative result does not actually disprove their data. Several technical reasons could plausibly explain the difference but there is also a relatively high probability (1 in 6) that, in two experiments with approximately the same beam dose, three events will be observed in one and none in the other. It is therefore useless to worry about severe technical failures; before those are seriously taken into account, statistical uncertainty needs to be reduced to a level no higher than 5% (1 in 20), which would necessitate a beam time ten times longer than the one we had used. We had already spent 24 days; nobody was prepared for another 220.

Progress here, retraction there

The year 2000 brought a time of waiting for the confirmation of element 118. Attempts were also made at the accelerator laboratories RIKEN in Saitama near Tokyo and GANIL in Caen, France, with no result. However, to be fair, the weight of these experiments is not very high with regard to a 'disprove' of the Berkeley result. Similar as the GSI experiment they suffer from insufficient statistical accuracy, despite the fact of uncertain capability to exactly reproduce the Berkeley experiment concerning choice of beam energy and detection efficiency. The negative results, now from three laboratories, created first doubts in element 118. The conclusion was that the best way to proceed would be a repetition at the BGS itself.

Meanwhile we performed at SHIP our confirmation experiments of element 112 in May 2000 and of element 111 in October. The positive results were already presented in Chapter 13.

In Dubna an irradiation started in summer 2000 to search for element 116. As a target they used ^{248}Cm which is one α particle heavier than the previously irradiated ^{244}Pu. The hope was, when using the same ^{48}Ca beam again, one could observe a 116 isotope which would α decay into the already studied nuclei ^{289}114 or ^{288}114. To find the daughter decays again is a necessary condition for the correctness of the measurement.

The experiment lasted 1.5 months. One α chain was observed on July 19th. The first α decay was measured with a lifetime of 47 milliseconds, the subsequent two α decays and spontaneous fission event agreed with the previously measured chain which had been assigned to ^{288}114. This was a quite convincing continuation of the experiments using the ^{48}Ca beam and a paper presenting the 'Observation of the decay of ^{292}116' was submitted to *Physical Review C* on October 2nd, 2000. The measured cross-section was 0.6 picobarn, not much less than for the synthesis of element 114, but a factor of 400 smaller than the limit we had reached together with the Berkeley people in our 1982–83 attempt using the same reaction (see Chapter 10). These numbers show how much the sensitivity was increased during the last 16 years, a factor of 400!

One event for element 116 was not enough. The cross-section was promisingly high for expecting more events in a follow-up experiment. In addition, there was still the open question of the 3 neutron channel. If also the very first chain observed in November 1998 and assigned to ^{289}114 could be measured as daughter after α decay of ^{293}116, then the Dubna people would have worked out a quite conclusive net made from decay chains and cross bombardments (these are cross-checks producing the same nuclei with different reactions). They continued to study 116 in November 2000.

The irradiation took 4.5 months in total. For many, many days there was no event. Then, on May 2nd, the next decay chain was measured and, surprisingly, already on May 8th the third one. The two more chains were in excellent agreement with the decay of ^{292}116 measured a year before, and with the α decays of ^{288}114, ^{284}112 and the spontaneous fission of ^{280}110. However, the 3 neutron evaporation channel was not observed.

The very positive progress made in experiments in Dubna and at the SHIP was in the shadow of the spectacular result on element 118 obtained in Berkeley. A first repetition at the BGS itself was performed at the end of 2000. However, due to a leakage in the apparatus the helium gas filling was not pure enough for a proper separation which resulted in an uncontrolled charge state of the reaction products. A second attempt was prepared for the end of April 2001.

At the beginning of April I met several of the Berkeley people at the annual conference of the American Chemical Society in San Diego. They were quite confident that now in the coming three week's run the 1999 results on their 118 element would be reproduced. The whole heavy element research world was looking forward.

At the time there were several smaller conferences on nuclear physics in Europe. The first was at the beginning of May in Sandanski, Bulgaria. Although invited, no people from Berkeley joined the conference. We understood this, because it was just half-time of their experiment. Oganessian reported about their second 116 decay chain from May 2nd measured in Dubna, I gave an overview talk on the status and future of superheavy element experiments and would have been very happy to include new results from Berkeley. But until the end of the conference on May 10th, no news.

The next conference was in the week from May 21st to 25th on Lipari island in Italy (physicists select nice places for their conferences). Again I had a review talk. The regular beam time in Berkeley must have been over, but still no definite news. There was some rumour that they have seen a decay chain, others knew that the new chain differed from the old ones and others had heard that there was nothing.

Usually the decay chains give a very clear signature for the decay of a heavy nucleus and they are relatively easy to detect from the amount of data using fast computers. It is especially easy when the chains are long and the lifetimes short. This was the case of the three chains of 118 measured in 1999. Something strange must have happened and we concluded that probably no decay chain was measured. Otherwise we would have heard the good news almost immediately, because this time no paper writing and priority reasons could cause a delay.

Two more conferences relevant to heavy element research took place in July, one in Finland and the other at the Baikal lake in Siberia. I was at both of these conferences, but nobody could make a definite statement on the result of the Berkeley experiment. However, the next major conference began on August 30th, 2001, in Berkeley itself. This was certainly the proper time to report on the status of the element 118 experiment. But astonishingly, there was no talk announced and no speaker from the heavy element group in the schedule.

I had to skip the Berkeley conference, because I had just came back from the Baikal conference on Sunday, August 29th. On Monday morning I looked through my email at the institute and was surprised. I read the following:

Sent: Friday, July 27, 2001 6:22 PM

Subject: Element 118

Dear Colleague,
I am in the unfortunate position of having to report some bad news. A very brief summary is included below in the text of a retraction (submitted July 26, 2001 to Physical Review Letters) of our element 118

publication from two years ago. We are greatly disturbed by this situation, and have been working very hard to understand what led to it. While we do not yet have the answers, we feel that at this time, it is important to alert the nuclear physics/nuclear chemistry community. We apologize for the erroneous report and any undesirable consequences that followed from it.

As for the new Kr + Pb experiments, we observed no element 118 events in our 86Kr + 208Pb experiments from this year, leading to a cross section limit of 2 pb . . . with limits as low as 0.7 pb in certain regions. This experiment was carried out under much better controlled conditions than our earlier experiments, and was actually several times more sensitive than the 1999 experiments. This new result will be written up and submitted in the coming weeks.

Ken Gregorich

. . .

Retraction of 'Observation of Superheavy Nuclei Produced in the Reaction of 86Kr with 208Pb'

In 1999, we reported [1] the synthesis of element 118 in the 208Pb(86Kr,n) reaction based upon the observation of three decay chains, each consisting of an implanted heavy atom and six sequential high-energy alpha decays, correlated in time and position. Prompted by the absence of similar decay chains in subsequent experiments [2–5], we (along with independent experts) re-analyzed the primary data files from our 1999 experiments. Based on these re-analyses, we conclude that the three reported chains are not in the 1999 data. We retract our published claim [1] for the synthesis of element 118.

[1] V. Ninov et al., Phys. Rev. Lett. 83, 1104 (1999).
[2] S. Hofmann and G. Münzenberg, Rev. Mod. Phys. 72, 733 (2000).
[3] K. Morimoto et al., AIP Conf. Proc. 561, 354 (2001).
[4] C. Stodel et al., AIP Conf. Proc. 561, 344 (2001).
[5] K. E. Gregorich et al., Phys. Rev. C (to be submitted).

I was shocked!

What could have been the reasons for such a dramatic action? I read the e-mail again and again but I could not find an answer. There was no clear reason given why the 1999 result was retracted. What astonished me was that Ken used phrases like 'We are greatly disturbed . . .' and 'We do not yet have the answers . . .'. Is this reason enough to retract a result from which the Berkeley people themselves were so much convinced two years ago and which most of us gave a good chance to be correct, although confirmation was demanded before the result could be generally accepted?

I was waiting until evening, then I tried to call Victor and Ken but could not reach them. A few days later, meanwhile, several journalists had already called and wanted more details. Just back from a holiday, Victor gave me some additional information. In summary it was that the result from the new experiment was negative and when they tried to re-analyse the old data, which had been stored on magnetic tape, severe problems arose to reconstruct the three decay chains from the data files. The reason for that was not understood.

I concluded that all the given arguments are not enough strong to disprove the 1999 result. If the Berkeley people do not have additional information which I do not know, then a retraction at this moment is unnecessary, it only creates confusion and uncertainty.

What can we learn from the Berkeley 118 story? What went wrong? I think that two unfortunate decisions were made two years ago. The first was to start such a challenging enterprise like the search for element 118 with a new and insufficiently tested apparatus. The second was, after the three highly interesting events were found in less then two weeks, to stop this experiment and prepare a search for still heavier elements. In German we have the proverb 'Der Spatz in der Hand ist besser als die Taube auf dem Dach' ('a bird in hand is worth two in the bush'). The Berkeley people had already 'three' in hand and let them fly off.

The above was written five months ago. Now, as we write, it is February 2002. There was no more news from Berkeley, but now we at SHIP were confronted with an unpleasant experience.

In December last year we had almost finished our paper on the confirmation of elements 111 and 112. It became a relatively long review paper (12 pages) on the new results which were already presented in Chapter 13. There was only one number open which we wanted to include in a figure for completeness. This was the position of the fifth α particle of the first 112 decay chain measured on February 1st, 1996. Because I could not find this value in my records, I asked Fritz to analyse it from the original files which are stored in a tape robot system in our computing centre. After a few hours he came back and told me that he easily could find the second chain which was measured on February 9th, 1996, but not the first chain. What has happened?

It took us several days to go through all of our old records and to make a complete new analysis of the 1996 run. Then we clearly had to realise that part of the output files of the 1996 analysis had been modified at the time. The modifications were so that additional feigned α particles were written in to the text files of the analysis to a measured background event. We easily could identify this event as a ^{212}Po nucleus which decayed by the measured 11.65 MeV α particle. Such nuclei are registered from time to time as background events. They are produced from the ^{208}Pb target by transfer reactions. However, due to the additional information feigning α particles, which are actually not in the original data files, a decay chain of five subsequent α decays was spuriously produced. It was assigned in 1996 to the first decay chain of element 112.

Two hectic weeks started. Knowing about one such incident, could there be more? We had to reanalyse all of our data measured since 1994 in order to be sure. This was a total of 34 decay chains, four of ^{269}110, eight of ^{270}110, thirteen of ^{271}110, six of ^{272}111, and the three of ^{277}112. Fritz carried out the main work of the reanalysis; people from our computing centre supported us. Finally we ended up with one more chain, the second chain of our first element 110 experiment performed in 1994, which had also been modified. In this case spurious signals feigning two α particles had been added to a background event so that a spurious chain was created. However, to our great contentment, all the other 32 decay chains were absolutely correct.

We rewrote our paper. Fortunately, the two 'lost' chains did not change our conclusions on the discovery of elements 110 and 112. We added a few paragraphs in the paper for explaining the results of our reanalysis, but the main part of the paper was dealing with our convincing, as we hope, new data from the confirmation experiments. The paper was sent on December 20th, 2001 to the *European Physical Journal* (the successor of *Zeitschrift für Physik*) for publication. A copy was sent to IUPAC for information. Happy about having finished our paper, but always thinking about why somebody has changed the data feigning two decay chains, we went into our Christmas holidays.

Where do we go from here?
For many years we were working hard to build up clear technical solutions for synthesising and identifying superheavy elements. The UNILAC had been built delivering stable and high current beams, the SHIP was constructed for separating the heavy nuclei from the beam, and the method was developed of detecting correlations in position and time of the decaying nuclei for establishing decay chains and unambiguous identification. Six new elements were the wages. We had learned much about their lifetime and how they can be produced.

Our data up to element 112 are generally accepted. The reason is that the decay chains are not located completely in the unknown but the lower part of the chains can be safely assigned to the decay data of well established nuclei. Yields behave systematically and inexplicable jumps were not measured. A continuation in small steps, element 113 as a next and 114 then, is a good strategy.

The Berkeley result on element 118 had to be retracted. However, the reaction of the krypton isotope with mass 86 and the closed neutron shell 50 with a target of the double magic ^{208}Pb is promising for further investigation.

A big step forward in the exploration of superheavy nuclei will have been taken by us with the recent discoveries on element 114 and 116 made in Dubna, if the data is eventually established. The measured yields for element 114 and 116 are not unreachably small and with the knowledge of what to look for, it should be possible to re-measure the data in Dubna itself, in our laboratory or in Berkeley, possibly also in the laboratories GANIL in France and RIKEN in Japan, where heavy element experiments are in preparation. Continuing on the path of hot fusion using the double magic ^{48}Ca beam and targets of californium (element 98) the superheavy element 118 may be produced. This cannot be done suddenly and certainly not in a forced attack, but within a few years we will have the result, I am sure.

Much work remains to explore the properties of the superheavy nuclei and the ways they are made. We need to know whether there is a region of particular stability at $Z = 114$, 120 or 126 or a more equally distribution across this range. As far as the neutron number is concerned, most workers do agree with $N = 184$ as a closed shell.

We are still a long way from element 126 but 114, 116 and 118 might be the route to an exploration of that group of superheavies. What would we need for that? Our equipment certainly has room for improvement. In order to approach the $N = 184$ neutron shell people think about working with more neutron rich, but then radioactive beams. Such beams will be available in some laboratories in the future. At GSI we require a more efficient accelerator for stable beams. The UNILAC is 30 years old: it was built to provide uranium beams but for exploring the superheavy elements we would have to have calcium, iron, zinc, germanium and krypton beams – perhaps even xenon or samarium. Relatively low energies (no higher than 7 MeV per nucleon) must be coupled to intense beam currents and continuous irradiation for months. That can be achieved only with a new superconducting linear accelerator specifically dedicated to exploring the superheavies.

Would it not all be too expensive? Is a study of the superheavy elements of any use? It won't be cheap but, since you have come this far in the book you will, I hope, have acquired some the sense of the excitement which has motivated workers in this field, so I will not try to convince you

with a host of detailed arguments that the outlay is worthwhile! To understand Nature we will always have to press on with our inquiries, searching
every nook and cranny for clues about what is going on now, what happened in the past and what is likely to take place in the future. This is so,
because we are curious.

15

Epilogue

I was extremely interested in nuclear physics even when I was at school. At the end of the 1950s the industrial countries had to solve their growing demand for energy with the help of nuclear power. 'One gram of uranium is the equivalent of a long freight train filled with coal!' was the message which excited me.

My other fascination was the atom, the smallest scrap of an element out of which all material and life is formed. The interaction of four tiny particles, the proton, neutron, electron and neutrino, provide all the variety of our universe – and maybe other universes that we don't know about too. The study of the bricks and all the variety of their apparent composition is a wondrous enterprise and, within this study, the search for superheavy elements has excited me for years.

It was a lucky chance for me that I was born at a time when atoms were already known, a microcosm of elementary particles which began to become visible at the end of the nineteenth century. In the last hundred years we have learned why the sun shines, how to make gold and how to use nuclear energy peacefully. Now we are beginning to manipulate life at the molecular level to help to cure a variety of diseases, to keep us fit and healthy to an ever more advanced age, and to reduce the hunger in the world.

But we also learned how to build atomic bombs and that human behaviour is not entirely reliable when it comes to controlling complicated technical facilities like nuclear power plants or, as we had to find out, the handling of data from a nuclear physics experiment. But much worse, even after the two terrible experiences in the first half of the twentieth century, devastating wars, even though they are local, have not yet be eliminated. Playing with fire shows up the dark side of human beings. Does it have to go on forever? Must it be just a dream; perhaps we can find a less belligerent map for mankind's journey into the future?

How well people can live together I learned from our worldwide community of heavy element researchers. Our common interest is the study of superheavy elements and progress in this field gave us contentment, independent of 'who' was lucky and 'where' the result was obtained. Errors cannot be completely excluded, therefore, to be critical is a demand, but to be unfair abject. In this spirit GSI was founded three decades ago. During that period I became acquainted with many outstanding people and excellent scientists from GSI itself and from all over the world. It is those people who create the stimulating atmosphere in which scientific research grows. Teamwork and the will for reaching the aim, is the fifth force which human beings add to the four governing the inanimate Nature. I am happy that I could live and work in such conditions and I am particularly grateful for the time which I could spent together with my colleagues in the SHIP group sailing towards superheavy island.

And finally, a last question which we ask at the beginning, during and at the end of our lives, and which Gottfried Benn expressed in his poem 'Nur zwei Dinge':

> *Durch so viele Formen geschritten,*
> *durch Ich und Wir und Du,*
> *doch alles blieb erlitten*
> *durch die ewige Frage: Wozu?*

'Only Two Suitcases'

Through so many forms to have passed,
through you and we and I,
yet endured it all at last
through the unceasing question: Why?

Probably we will not find the answer. However, the efforts we are making to find it bring sense into our life, our aiming, our wishing, our dreaming and our acting.

Bibliography

The following were used in the preparation of this book:

Barber, R.C., Greenwood, N.N., Hrynkiewicz, A.Z., Jeannin, Y.P., Lefort, M., Sakai, M., Ulehla, I., Wapstra, A.H. and Wilkinson, D.H. Discovery of the transfermium elements, in *Progress in Particle and Nuclear Physics*, volume 29, pages 453–530 (1992).

Blackett, P.M.S. and Lea, D. Impact of alpha particle with nitrogen nucleus, in *Proceedings of the Royal Society of London*, volume A 136, pages 325–8 (1932).

Crandall, J.L. Synthesis of the transuranium nuclides, in *Gmelin Handbuch der Anorganischen Chemie*, volume 7b, Transurane, pages 1–121, Springer Verlag, Berlin-Heidelberg-New York (1974).

Flerov, G.N. Map of isotopes, in *Physica Scripta*, volume 10A, page 1 (1974).

Flerov, G.N. and Ter-Akopian, G.M. Synthesis and study of atomic nuclei with Z>100, in *Progress in Particle and Nuclear Physics*, volume 19, pages 197–239 (1987).

Ghiorso, A., Nitschke, J.M., Alonso, J.R., Alonso, C.T., Nurmia, M., Seaborg, G.T., Hulet, E.K. and Lougheed, R.W. Element 106, in *Physical Review Letters*, volume 33, pages 1490–3 (1974).

Herrmann, G. Synthesis of the heaviest chemical elements – results and perspectives, in *Angewandte Chemie*, International Edition, volume 27, pages 1417–36 (1988).

Herrmann, G. The discovery of nuclear fission – good solid chemistry got things on the right track, in *Radiochimica Acta*, volume 70/71, pages 51–67 (1995).

Hofmann, S. New elements – Approaching Z = 114, in *Reports of Progress in Physics*, volume 61, pages 639–89 (1998).

Hofmann, S. and Münzenberg, G. The discovery of the heaviest elements, in *Reviews of Modern Physics*, volume 72, pages 733–67 (2000).

Hofmann, S., Reisdorf, W., Münzenberg, G., Heßberger, F.P., Schneider, J.R.H. and Armbruster, P. Proton radioactivity of 151Lu, in *Yeitschrigt fürPhysik A*, volume 305, pages 111–23 (1982).

Leachman, R.B. Nuclear fission, in *Scientific American*, pages 49–59 (1965).

Münzenberg, G. and Schädel, M. *Moderne Alchemie – Die Jagd nach den schwersten Elementen*. Friedr. Vieweg und Sohn Verlagsgesellschaft, Braunschweig, pages 1–258 (1996).

Ninov, V., Gregorich, K.E. and McGrath, C.A. The Berkeley gas-filled separator, *Proceedings of the International Conference on Exotic Nuclei and Atomic Masses, ENAM 98*, American Institute of Physics, USA, pages 704–7 (1998).

Oganessian, Y.T. Fusion and fission induced by heavy ions. *Lecture Notes in Physics*, volume 33, pages 221–52, (1975).

Oganessian, Y.T., Demin, A.G., Danilov, N.A., Flerov, G.N., Ivanov, M.P., Iljinov, A.S., Kolesnikov, N.N., Markov, B.N., Plotko, V.M. and Tretyakova, S.P. On spontaneous fission of neutron-deficient isotopes of elements 103, 105 and 107, in *Journal Nuclear Physics A*, volume 276, pages 505–22 (1976).

Oganessian, Y.T., Demin, A.G., Iljinov, A.S., Tretyakova, S.P., Pleve, A.A., Penionzhkevich, Y.E., Ivanov, M.P. and Tretyakov, Y.P. Experiments on the synthesis of neutron-deficient kurchatovium isotopes in reactions induced by 50Ti ions, in *Nuclear Physics A*, volume 239, pages 157–71 (1975).

Oganessian, Y.T., Hussonnois, M., Demin, A.G., Kharitonov, Y.P., Bruchertseifer, H., Constantinescu, O., Korotkin, Y.S., Tretyakova, S.P., Utyonkov, V.K., Shirokovsky, I.V. and Estevez, J. Experimental studies of the formation and radioactive decay of isotopes with $Z = 104$–109, in *Radiochimica Acta*, volume 37, pages 113-20 (1984).

Oganessian, Y.T., Tretyakov, Y.P., Iljinov, A.S., Demin, A.G., Pleve, A.A., Tretyakova, S.P., Plotko, V.M., Ivanov, M.P., Danilov, N.A., Korotkin, Y.S. and Flerov, G.N. Synthesis of neutron-deficient isotopes of fermium, kurchatovium, and element 106, in *Journal of Experimental and Theoretical Physics Letters*, volume 20, pages 265–6 (1974).

Oganessian, Y.T., Utyonkov, V.K. and Moody, K.J. Voyage to Superheavy Island, in *Scientific American*, pages 63–7 (2000).

President of the Technical University Berlin (ed.) *Die Geschichte der Entdeckung der Kernspaltung* (Catalog of the exhibition: The history of the discovery of nuclear fission). Library of the University Berlin, Berlin, Germany (1989).

Schechter, B. The short bright life of element 109, in *Discover*, pages 98–106 (1982).

Seaborg, G.T. Transuranium elements: the synthetic actinides, in *Radiochimica Acta*, volume 71/72, pages 69–90 (1995).

Seaborg, G.T. and Bloom, J.L. The synthetic elements: IV, in *Scientific American*, pages 57–67 (1969).

Seaborg, G.T. and Loveland, W.D. *The elements beyond uranium*. John Wiley and Sons, Inc., New York (1990).

Segre, E. *Nuclei and Particles* Second edition. W.A. Benjamin, Inc., New York, Amsterdam (1965).

Sobiczewski, A. Review of recent SHE predictions, in *Physica Scripta*, volume 10A, pages 47–52 (1974).

Wideröe, R. (1928) cited in *History of Linacs* (AccSys Technology) http://www.accsys.com (accessed 26 March 2000).

Zafiratos, C.D. The texture of the nuclear surface, in *Scientific American*, pages 100–8 (1972).

Gottfried Benn's poem 'Nur zwei Dinge' was paraphrased by Marcel Bayer in his 'Only Two Suitcases' in *Poetry*, Edition October–November 1998, Vol. 173, Modern Poetry Association, Chicago.

Name index

Subject index

Element index

Printed and bound by CPI Group (UK) Ltd, Croydon, CR0 4YY

01/11/2024

01782625-0002